The Energy Transition in Japan

This book offers a distinctive and comprehensive view of the energy transition of Japan, with a particular focus on the rise of smart cities.

Drawing on real examples from Japan's journey towards carbon neutrality, this volume examines a variety of topics ranging from laws and policies to technological and managerial solutions, discussing them in the context of Japan's energy transition. Among the issues covered by the book are climate action planning, sustainable waste management, energy poverty, decarbonisation, e-methane, transport policies, and smart grids. The book also explores the regulatory tools that either support or hinder the development of smart cities in Japan, and how Japan can leverage its national solutions globally. In this way, this book serves as a guide for global climate action and energy transitions around the world.

Focusing on both stories of success and lessons learned, this book will be a valuable resource for students and scholars of energy transitions, climate action, smart cities, as well as Asian and Japanese studies more broadly.

Maciej M. Sokołowski, PhD, DSc, is a Specially Appointed Associate Professor at the Faculty of Policy Management of Keio University, also affiliated with the Faculty of Law and Administration at the University of Warsaw. Professor Sokołowski has extensive experience in energy law and the energy sector; he has authored 100 papers and reports, including three solo books on energy regulation, combined heat and power, and the energy transition. Professor Sokołowski is a fellow of several institutions and networks, including the Sustainability College Bruges, the SI Network for Future Global Leaders, the Polish Electricity Association, the Australian Network for Japanese Law, the Japan Association of EU Studies, and the Japan Society of Public Utility Economics. Professor Sokołowski is also a lead author of the Intergovernmental Panel on Climate Change (IPCC) Special Report on Climate Change and Cities, being responsible for Chapter 4: "How to Facilitate and Accelerate Change". He has been awarded numerous distinctions, including the Swiss Government Excellence Scholarship, the Swedish Institute Visby Programme scholarship, and the Prime Minister of Poland's Research Award. In 2024, Professor Sokołowski was named one of Stanford University's "World's Top 2% Scientists".

Fumio Shimpo, PhD, is Professor of Law at the Faculty of Policy Management of Keio University. Professor Shimpo is an active scholar in the fields of data protection, privacy, information law, AI, and robot law in Japan. He serves as the Chairperson of the Board of Directors of the Association of Law and Information Systems, the Executive Director of the Japanese Constitutional Law Society, a Board Member of the Japan Society of Information and Communication Research, the Director of the Law and Computer Society, and a Senior Research Fellow at the Institute for Information and Communications Policy of the Ministry of Internal Affairs and Communications. He was previously the Commissioner for International Academic Exchange at the Personal Information Protection Commission of Japan (2018–2023) and the former Vice-Chair of the OECD Working Party on Security and Privacy in the Digital Economy (2009–2016).

Routledge Explorations in Energy Studies

Northern Indigenous Community-Led Disaster Management and Sustainable Energy
Ranjan Datta, Margot Hurlbert and William Marion

Energy Policy Design in the South-Eastern Mediterranean Basin
A Roadmap to Energy Efficiency
Bertug Ozarisoy and Hasim Altan

Living with Energy Poverty
Perspectives from the Global North and South
Edited by Paola Velasco-Herrejón, Breffní Lennon and Niall Dunphy

Transitioning Fossil-Based Economies
Sustainable Strategies for Energy Change
Hassan Qudrat-Ullah

The Future of Liquified Natural Gas in a Decarbonising World
Omran Al-Kuwari

Energy Use in Bitcoin Mining
The Environmental Impact of Cryptocurrencies
Benjamin A. Jones; Andrew Goodkind and Robert P. Berrens

Smart Cities and Japan's Energy Transition
Past, Present, and Future
Edited by Maciej M. Sokołowski and Fumio Shimpo

The Energy Transition in Japan
Smart Cities and Smart Solutions
Edited by Maciej M. Sokołowski and Fumio Shimpo

For more information about this series, please visit: www.routledge.com/Routledge-Explorations-in-Energy-Studies/book-series/REENS

The Energy Transition in Japan

Smart Cities and Smart Solutions

Edited by Maciej M. Sokołowski and Fumio Shimpo

Routledge
Taylor & Francis Group
LONDON AND NEW YORK

First published 2025
by Routledge
4 Park Square, Milton Park, Abingdon, Oxon OX14 4RN

and by Routledge
605 Third Avenue, New York, NY 10158

Routledge is an imprint of the Taylor & Francis Group, an informa business

British Library Cataloguing-in-Publication Data
A catalogue record for this book is available from the British Library

ISBN: 978-1-032-74896-2 (hbk)
ISBN: 978-1-032-74899-3 (pbk)
ISBN: 978-1-003-47141-7 (ebk)

DOI: 10.4324/9781003471417

Typeset in Times New Roman
by Newgen Publishing UK

Contents

Contributors

Yasuhiro Hayashi, PhD, is Professor in the Department of Electrical Engineering and Bioscience at Waseda University, where he is also the Chairman of the Advanced Collaborative Research Organization for Smart Society and Director of the Research Institute for Advanced Network Technology. His research focuses on power systems, smart grids, energy management systems, and demand response, with numerous publications in these areas.

Hideo Ishii, PhD, is Professor and the Secretary General of the Advanced Collaborative Organization for Smart Society at Waseda University. Professor Ishii has been conducting research in electric power systems, particularly focusing on the integration of renewable energy and the application of demand response and virtual power plants in power system management. He was a member of the Japanese national committee tasked with determining the specifications for the next generation of smart meters. He has authored over 100 academic papers and articles and is a frequent speaker at international conferences. Since 2020, he has been the Chair of the International Electrotechnical Commission TC8 SC 8C.

Hiroshi Ito, PhD, is Professor of International and Sustainable Development and Education. He currently teaches at Nagoya University of Commerce and Business (NUCB). Before joining NUCB, he served as an education expert with UNESCO (France), UNICEF (Philippines), and JICA (Ecuador and Paraguay). His research interests lie in educational assessment and environmental policy, and he has several publications in these fields.

Misako Kachi is Senior Researcher at the Earth Observation Research Center of the Japan Aerospace Exploration Agency (JAXA). She graduated from Hokkaido University with a degree in Earth and Planetary Physics and later from the University of Tokyo with a degree in Planetary Science. Her research interests include microwave remote sensing and its applications. She serves as the Manager for GCOM-W & AMSR3 Research and Marine Environment Monitoring at EORC and is a member of the Japan Geophysical Union and the Meteorological Society of Japan. She was the recipient of prizes for science and technology in the Commendation for Science and Technology by the Minister of Education, Culture, Sports, Science and Technology in 2016 and 2021.

Justyna Kamila Kanas is a researcher in Energy and Climate Law. She graduated with honours from the Faculty of Law and Administration at the University of Warsaw and is preparing her doctoral thesis on hydrogen regulations. She is the author of several papers and technical reports concerning energy transition and a multiple scholarship recipient for her academic achievements and participation in interdisciplinary energy programmes.

Kazuhiko Kato, PhD, is Professor at the Graduate School of Management, Nagoya University of Commerce and Business (NUCB). Dr Kato has lived in Australia, where, after completing his education at Bond University and the University of Sydney, he served as a senior manager of a local subsidiary. There, he was involved in planning domestic sales strategies for a venture company that emerged during the liberalization of telecommunications. His research interests cover strategic management, international business, and digital transformation.

Takuji Kubota, PhD, is a senior researcher at the Earth Observation Research Center of the Japan Aerospace Exploration Agency (JAXA). His research interests are in algorithm development for spaceborne radar and microwave radiometers for the TRMM, GPM, and EarthCARE missions. Dr Kubota is a member of the Remote Sensing Society of Japan, Meteorological Society of Japan, Japan Geoscience Union, American Meteorological Society, and American Geophysical Union. He was the recipient of prizes for science and technology in the Commendation for Science and Technology by the Minister of Education, Culture, Sports, Science and Technology in 2016, and the Gambo-Tatehira Award by the Meteorological Society of Japan in 2019.

Satoshi Kurokawa, PhD, is Professor of Environmental Law and Administrative Law at Waseda University, School of Social Sciences. Professor Kurokawa has published in leading academic journals in Japan. His research focuses on domestic environmental regulation by governments, climate change law, and energy law, including regulations on renewable energy and nuclear energy. He has been an editor for the *Journal of Environmental Law* in Japan.

Shinichi Kusanagi, SJD, is Professor of Administrative Law at the University of Hyogo, School of Economics and Management. His research focuses on energy law and environmental law. Professor Kusanagi has published in leading academic journals and is the author of several books, including his recent work on smart energy networks. He is also a frequent speaker at academic conferences and has received numerous awards for his contributions to the field.

Xiaoyue Liu, PhD, is Assistant Professor of Waste Management and Recycling Policy at Tohoku University. Her research focuses on sustainable waste management and efficient recycling systems, with several publications in these fields. She is also conducting international cooperative research in developing countries to investigate issues of marine plastic pollution in the Philippines and energy poverty in Mongolia.

Gaku Manago, PhD, is Project Lecturer in Environmental Sociology and International Cooperation Studies. He has been researching sustainable recycling methodologies and water supply sanitation at Tohoku University. His research interests include not only environmental studies and sociology but also optical engineering; he conducts cross-disciplinary works and collaborates with domestic and international companies. Additionally, he is currently working on establishing a startup company – the Innovation Leaders Summit has selected him as one of the Top 100 Attractive Start-ups.

Naoko Matsuo is Senior Engineer at the Satellite Applications and Operations Center, Space Technology Directorate I, of the Japan Aerospace Exploration Agency (JAXA). She has been involved in international cooperation projects for Earth observation satellites and the launch of future satellites. She heads the secretariat of the Consortium for Satellite Earth Observation (CONSEO), which works on formulating satellite strategies through industry–academia–government collaboration.

Piotr Mikusek is Analyst of Climate and Energy Regulations who has been teaching at the University of Warsaw, the Faculty of Law and Administration. He holds a Master's degree in law and philosophy and has published extensively in leading academic journals. His research focuses on regulatory tools used to achieve decarbonisation. He participates in expert groups working alongside the Polish government administration to support the energy transition.

Shilpi Mittal, PhD, is a researcher in Urban Studies. Dr Mittal has published extensively in leading academic journals and edited book volumes. Her research focuses on the comprehensive assessment of urban development, sustainable habitats, and quality-of-life issues in rapidly transforming societies. Dr Mittal is a recipient of the Ministry of Education, India's GATE Scholarship (2005), and a Visiting Scholarship (2020) at the University of Glasgow, Scotland.

Shiori Osanai, PhD, is a researcher in Educational Sociology and an Academic Research Fellow at the Graduate School of International Cultural Studies at Tohoku University. Her research focuses on building a new educational model for developing human resources who will contribute to regional revitalisation. She also conducts comparative research between Japan and South Korea on policies and educational approaches for regional revitalisation, focusing not only on Japan but also on ageing and depopulated regions in South Korea. As part of her contributions to the local community, she is also continuously involved in sustainability education at primary and secondary schools in areas affected by the Great East Japan Earthquake.

Jacek Piecha, PhD, is Assistant Professor at the Faculty of Law and Administration, University of Warsaw. He is the author of several articles concerning the theory of administrative law and administrative proceedings. Currently, his research focuses on the legal aspects of applying new technologies in public administration and their influence on human rights.

Mahendra Sethi, PhD, is an urban environment expert exploring the role of cities at the interface of global changes. As a senior fellow, he leads the urban sustainability group at the Indian Society of Applied Research & Development and the ICLAP 2050 project, supported by the Asia Pacific Network for Global Change Research. Dr Sethi has published over 45 peer-reviewed research papers, chapters, and policy articles. His research focuses on the sustainable built environment, megacities, policy and governance, pandemic resilience, environmental planning, optimisation models, green technologies, and responsible development in ecologically sensitive areas. He is a recipient of the Alexander von Humboldt Fellowship at TU Berlin, the UNU-IAS PhD Fellowship at United Nations University, Japan, and a visiting scholarship at Kyoto University as well as at the Ministry of Education, India's GATE Scholarship.

Fumio Shimpo, PhD, is Professor of Law at the Faculty of Policy Management of Keio University. Professor Shimpo is an active scholar in the fields of data protection, privacy, information law, AI, and robot law in Japan. He serves as the Chairperson of the Board of Directors of the Association of Law and Information Systems, the Executive Director of the Japanese Constitutional Law Society, a Board Member of the Japan Society of Information and Communication Research, the Director of the Law and Computer Society, and a senior research fellow at the Institute for Information and Communications Policy of the Ministry of Internal Affairs and Communications. He was previously the Commissioner for International Academic Exchange at the Personal Information Protection Commission of Japan (2018–2023) and the former Vice-Chair of the OECD Working Party on Security and Privacy in the Digital Economy (2009–2016).

Maciej M. Sokołowski, PhD, DSc, is a Specially Appointed Associate Professor at the Faculty of Policy Management of Keio University, also affiliated with the Faculty of Law and Administration at the University of Warsaw. Professor Sokołowski has extensive experience in energy law and the energy sector; he has authored 100 papers and reports, including three solo books on energy regulation, combined heat and power, and the energy transition. Professor Sokołowski is a fellow of several institutions and networks, including the Sustainability College Bruges, the SI Network for Future Global Leaders, the Polish Electricity Association, the Australian Network for Japanese Law, the Japan Association of EU Studies, and the Japan Society of Public Utility Economics. Professor Sokołowski is also a lead author of the Intergovernmental Panel on Climate Change (IPCC) Special Report on Climate Change and Cities, being responsible for Chapter 4: "How to Facilitate and Accelerate Change". He has been awarded numerous distinctions, including the Swiss Government Excellence Scholarship, the Swedish Institute Visby Programme scholarship, and the Prime Minister of Poland's Research Award. In 2024, Professor Sokołowski was named onc of Stanford University's "World's Top 2% Scientists".

Naoko Sugita, PhD, with a dissertation in Public Policy Analysis. She has been leading efforts to foster innovation in the space sector by connecting academia, industry, and government at the Japan Aerospace Exploration Agency (JAXA). Dr Sugita is interested in both research and practicing innovative space projects and studies. Among the projects she initiated was a collaboration with a startup and local government, which later won the Prime Minister's Japan Open Innovation Prize. She is a frequent speaker at international conferences and has given numerous lectures at academic institutions.

Aki Suwa, PhD, is Professor at the Faculty of Contemporary Society, Kyoto Women's University. She conducts research on environmental and sustainable policy approaches, based on community and regional analysis in Japan and across Asia. Prior to her appointment, she was a Research Fellow at the United Nations University Institute of Advanced Studies, under the Sustainable Urban Futures Programme, working on a project funded by the Ministry of the Environment, Japan, on urban development with co-benefits approach. She has a variety of academic and business experiences, maintaining close collaborative relations with non-governmental climate research institutes in Japan.

Tadao Tanabe, PhD, is Professor of Engineering Design. He has been teaching at Shibaura Institute of Technology. Tanabe has published over 100 papers in a wide range of leading journals on optics, semiconductors, and equipment development. His research interests include terahertz technology as a basis for recycled design of waste plastics and building infrastructure, and he has received numerous awards for his contributions, especially in the field of materials.

Madeline Taylor, PhD, is Associate Professor of Energy Law and Director of Research Training at Macquarie Law School, Co-Lead at the Transforming Energy Markets Research Centre, an Australian Research Council Industry Fellow, and a Honorary Associate at the Sydney Environment Institute, at the University of Sydney. Her award-winning research focuses on navigating complex energy regulatory systems to develop legal instruments that support the effective, sustainable, and justice-driven scaling up of innovative energy technologies. She has published extensively in leading journals. Her latest co-edited book focuses on drawing legal lessons from traditional offshore energy sectors to encourage best-practice regulation in offshore energy net zero industries.

Jun Yamashita, PhD, is Professor of Geography and has taught at the Faculty of Social and Cultural Studies, Kyushu University. He has published in leading academic journals and contributed to several influential books. His research focuses on urban transitions and sustainable urban development. Professor Yamashita is a frequent speaker at international conferences and has served as an expert for international organisations.

Jeongsoo Yu, PhD, is Professor of Waste Management and Sustainable Recycling. He has been teaching at the Graduate School of International Cultural Studies at Tohoku University. Professor Yu has published extensively in leading academic journals. He also conducts several international collaborative research projects, including his recent work on development strategies and policy trends of next-generation vehicle batteries in China, Japan, and South Korea. Recently, he developed a plastic sorting machine using terahertz waves, aimed at achieving a circular economy based on sustainable resource recycling. Professor Yu has been managing the Asian Automotive Environmental Forum since 2009 and has delivered education on the Sustainable Development Goals to areas affected by the Great East Japan Earthquake, working with various stakeholders.

Piotr Zieliński is an attorney-at-law and architect specialising in public law. As a scholarship holder of the Japanese government, he completed Master's studies at the National Graduate Institute for Policy Studies (GRIPS) in Tokyo and the GRIPS Alliance Diplomacy Academy. He gained professional experience in the private sector and the Polish administration. He is a participant in the SYNERGIA programme, a network for cooperation and exchange of experiences among high-level officials from Central and Eastern Europe.

Foreword

Energy-related issues are always considered as a critically important problem and challenge for Japan in her modern history. Japan, as an advanced industrial country with over 100 million population, needs vast amount of energy to fuel its economy and support civic life of Japanese citizens. Currently, over 80% of Japan's primary energy requirement is supplied by fossil fuels namely, oil, natural gas, and coal. Because of the lack of domestic fossil fuel resources, Japan is heavily dependent on energy imports including those from the Middle East. It is noteworthy that Japan's dependence on Middle East in her crude oil supply is now as high as 95%. High energy import dependence and high reliance on fossil fuels are the major causes of the serious concern over Japan's energy security vulnerability and climate change protection.

Since the 1st Oil Crisis in the 1970s, Japan has continued to make national efforts to enhance energy security. Recent experiences of serious instability in global energy market caused by the Russian invasion of Ukraine resulted in renewed highlight on the importance of energy security in Japan. At the same time, Japan has committed to implement her national plan for deep greenhouse gas (GHG) emission reduction in a form of Nationally Determined Contribution (NDC) submitted to United Nations. In other words, Japan is now in the midst of a long-term journey for energy transition in which simultaneous achievement of energy security and deep GHG emission reduction is being pursued.

In this context, this book is an extremely timely and interesting attempt with numerous useful analyses in respective chapters. The coverage of the research topics in the book is well diversified and all relevant to discuss key challenges for Japan's energy transition. As is found in the book, how to successfully conduct and implement "smart energy transition" is really a key to the successful survival for Japan. I believe the readers can find the book very useful and valuable for the better understanding of current status and future prospect of Japan's energy transition to achieve energy security and decarbonisation simultaneously.

Ken Koyama
Senior managing director and Chief economist
Institute of Energy Economics, Japan

Preface

Japan has always been an interesting benchmark for the world. This applies to Japanese culture, technologies, or economy. This holds true for the energy sector and smart cities, which offer numerous solutions that can be applied globally.

To address these areas, our book gathers a variety of topics ranging from laws and policies to technological and managerial solutions, to discuss them in the context of Japan's energy transition. Among the issues covered by this book are solutions for climate action planning, sustainable waste treatment, mobility as a service, spatial planning, data management, e-methane, or smart grids. The book also explores the regulatory tools that support or hinder the development of smart cities in Japan, and how Japan can leverage its national solutions globally.

Based on real examples from Japan's journey towards carbon neutrality, this book offers a distinctive and comprehensive view on smart cities in the energy transition of Japan. It is a helpful reference for anyone who seeks to learn more about the current status and future outlook of smart cities in Japan – but given the Japan's influential position for the world – their impact on global energy transformation. In this way, this book serves as a guide for global climate action in smart cities and energy transition based on smart solution around the world.

Maciej M. Sokołowski
Fumio Shimpo

Acknowledgements

First and foremost, we would like to thank all authors who were willing to work with us on this wonderful project. We are very grateful for your contributions and look forward to furthering cooperation. We also thank Dr Ken Koyama, Senior Managing Director and Chief Economist of the Institute of Energy Economics, Japan, for writing his Foreword to our book.

Special thanks go to Ms Justyna Kamila Kanas, who helped us gather useful reference materials needed for the research basis of this project, as well as to Ms Klaudia Pryjmak, who served as assistant editor, offering tremendous help in the coordination process and the first proofreading of each chapter. You really did it well.

We also thank all editors from Routledge engaged in publishing this book – you did a great job mastering our work. It is wonderful that we again had this pleasure of working together.

Last but not least, we would like to thank all colleagues engaged in the JST Moonshot R&D project, JPMJMS2215, including our team member, Ms Kimie Hatakeyama. Your perfect support and assistance gave us precious time so much needed for editing this book.

本書が多くの方々の支えと協力によって完成したことに改めて感謝申し上げます。

この場を借りて、心より深い感謝の意を表します。

Maciej M. Sokołowski
Fumio Shimpo

1 Searching for Smart Solutions for Energy Transition in Japan

Can One Find Them in Japanese Cities?

Maciej M. Sokołowski and Fumio Shimpo

1.1 Introduction

The energy transition has emerged as one of the most important priorities for the world. Driven by climate-related problems and the Sustainable Development Goals (SDGs), it is a long-term challenge that countries across the globe have been trying to address. Although generally heading in the same direction, different solutions for the energy transition have been offered worldwide. Among these, we must choose wisely – we must choose smartly. This, in a natural way, evokes a question: which solutions are smart?

Given the promise of smart solutions, it is logical to explore the potential of smart cities. Smart cities are very good platform for facilitating the smart revolution that we have been facing, ranging from smartphone, via smart transportation, to smart grid and smart house (see Visvizi and Troisi, 2022; Sokołowski and Visvizi, 2023). With more than half of the world's population living in cities, urban complexes are those agents of change – with their citizens driving the shift towards smart solutions. It is also about resources available there, including workplaces and education facilities that attract a particular type of people.

Cities, while offering numerous opportunities, also present significant challenges. The rural-to-urban migration of young individuals leads to a depletion of human capital in rural areas. The expanding urban population necessitates substantial investments in infrastructure, exacerbating housing costs. Furthermore, these urban centres face environmental and climate pressures stemming from traffic, pollution, waste, and water management, as well as energy services. How we address these city-related issues will determine whether humanity can transition to sustainability (see Georgescu *et al.*, 2021).

To tackle these challenges, several solutions have been offered. In this book, we discuss some of them in the context of Japan's energy transition, once described in the following words by Prime Minister Yasuo Fukuda:

> in order to shift Japan to a low-carbon society, in the near future, we will undertake a fundamental rethinking of all our societal systems, including our

DOI: 10.4324/9781003471417-1

> production systems, our lifestyles, and the state of our cities and transportation. We will seek to expand this low-carbon society both at home and abroad and play a leading role in transforming the globe into a low-carbon planet.
>
> (Fukuda, 2008)

As global urbanisation continues to rise, Japan's solution-oriented approach to smart cities can serve as a compelling benchmark, particularly in the context of energy transition. This encompasses technological advancements in renewables, hydrogen, batteries, or artificial intelligence (AI) and robotics, which are gaining traction in Japan (see Sokołowski, 2022), particularly in Japanese cities that have a long history of leading the country's transformation. Driven by recent policy goals, including reduction of greenhouse gas (GHG) emissions to net zero by 2050, aiming for a carbon-neutral and decarbonised society, Japanese cities are also becoming more climate-friendly locations for energy transition (see Sokołowski, 2022), investing in renewable energy installations, such as rooftop solar panels (photovoltaics), or reducing energy consumption. However, considerable efforts are still needed, especially in the areas of buildings and energy efficiency, as well as electrification of transportation.

1.2 Book's Structure and Issues Discussed

The issues analysed in this book are divided into 13 chapters. Following this introduction, Chapter 2 highlights Japan's policy goals related to climate change, along with visions presented by Japanese prime ministers at different international forums. As many of these proposals have been focused on technological innovation to address global environmental and climate problems, we link this approach to *monozukuri*, Japan's deep-rooted commitment to manufacturing excellence by examining public-driven programmes that have stimulated Japanese research and development capacity in the energy sector, as well as other policy incentives focused on technological innovations to address climate change. Furthermore, we discuss energy efficiency as a key component of Japan's technology-oriented approach. By combining these two elements (energy efficiency and the focus on technology), we analyse their role in energy transition of Japan, with smart cities at its core.

In Chapter 3, Mahendra Sethi, Shilpi Mittal, and Aki Suwa explore integrated climate action planning (ICLAP 2050) tool, discussing its application in Japanese cities. The ICLAP tool combines data and information for climate mitigation and adaptation, providing evidence-based recommendations, and integrating database, evaluation, and simulation features. The tool downscales global and regional climate scenarios, conducts meta-analysis of cities to forecast demographics, economy, energy, and GHG emissions. It also conducts a systematic review of climate interventions from city case studies worldwide. In this context, Chapter 3 presents the results of applying the ICLAP tool in two Japanese cities – Tokyo and Osaka – to offer data-driven simulations for local governments to support them in implementing effective climate actions.

Chapter 4, authored by Hideo Ishii and Yasuhiro Hayashi, highlights the 5Grids concept, which pertains to data collection within the five core elements of urban infrastructure: electricity, gas, water supply, transport, and communication. These five sectors are analysed in terms of data collection, data management, and data integration with external systems. This chapter discusses them as components of Japan's energy transition, offering a framework that enables quantitative data analysis to support effective policies for a zero-carbon, decentralised, and digitalised society.

In Chapter 5, Jeongsoo Yu, Xiaoyue Liu, Tadao Tanabe, Gaku Manago, and Shiori Osanai provide an overview of the history and challenges of waste management in Japan. They address the current state and issues of waste treatment and recycling, including energy recovery, focusing on several case studies of innovative technologies for effective plastic recycling. This chapter also explores cooperative recycling behaviours and social engagement in SDGs education to offer guidance on how to achieve a sustainable recycling network with cross-sectoral collaboration. Finally, the authors propose policy recommendations for creating a smart city based on a carbon-neutral and circular economy.

Chapter 6, written by Shinichi Kusanagi, covers creating a society with nearly zero GHG emissions with the help of synthetic methane produced from renewable or clean resources. To highlight the background of this process, this chapter examines the Japanese policy framework for e-methane production as a promising technology for making clean city gas, crucial for achieving a zero-carbon society. The chapter also reviews various possible policy responses to achieve this and promote Japan's Green Transformation (GX).

Chapter 7, authored by Jun Yamashita, offers a discussion on the net-zero goal in the Japanese transport sector. By comparing Mobility as a Service (MaaS) project implemented in smart cities in Japan, Europe, and the United States, he addresses the potential of MaaS as a policy instrument to achieve net-zero emissions. Moreover, this chapter identifies socio-economic, legal, and other challenges faced by Japanese MaaS projects in the context of energy transition in the transport sector and provides policy implications for these projects.

In Chapter 8, Jacek Piecha, Justyna Kamila Kanas, and Piotr Mikusek examine the role of central and local authorities in the implementation of tools that foster and develop smart cities in Japan. Through analyses of Japanese examples, they illustrate how the adoption of smart city solutions can prompt the engagement of public administrations in these innovative initiatives. This engagement leads to the creation of solutions that transcend the conventional relationship between governing bodies and governed entities. As discussed in the chapter, enhancing civic engagement and channelling private investment towards public goals like those related to climate and energy, improves the efficiency of local governance.

In Chapter 9, Piotr Zieliński contrasts spatial planning in Japan and the European Union (EU) with respect to smart sustainable solutions. His chapter explores what defines smart cities in Japan and the European Union and compares the approaches adopted to achieve the SDGs. To address these, the chapter identifies relevant Japanese and European laws and analyses official strategies to assess the future directions of efforts aimed at integrating urban technological advancements with the values and assets that merit legal protection.

Chapter 10, contributed by Madeline Taylor, examines two critical energy policy and regulatory developments in Australia that are likely to influence Japan's smart city agenda. First, it explores Australia's advancements in blending hydrogen into gas networks. Second, it highlights the significance of Australia's rooftop solar photovoltaics and prosumer policies in creating future marketplaces for peer-to-peer trading. These two valuable case studies provide valuable insights into pathways for fuel and electricity decarbonisation, serving as interesting benchmarks for future Japanese smart city initiatives.

In Chapter 11, Satoshi Kurokawa investigates the issue of energy poverty through the lens of energy transition in smart cities. He analyses Japanese legal acts and programmes aimed at facilitating the transition to a low-carbon society, reviewing their impact on households and electricity prices. This chapter also defines how the problem of energy poverty is understood in Japan and how it has been studied in research. Moreover, the chapter presents smart city solutions that aim to mitigate issues related to energy poverty.

Chapter 12, authored by Hiroshi Ito and Kazuhiko Kato, offers a discussion on the concept of the corporate smart city developed by Toyota Motor Corporation: Woven City, a smart city currently under construction in Shizuoka prefecture. While Woven City presents a notable case study in several aspects, comprehensive and publicly accessible information about the project is limited. Therefore, the chapter examines the nature and characteristics of Woven City in comparison with other corporate smart cities, through document analysis and interviews with relevant stakeholders.

Finally, Chapter 13, written by Naoko Sugita, Naoko Matsuo, Takuji Kubota, and Misako Kachi addresses the possible contribution of satellite Earth Observation to Japan's energy transition. In this chapter, the authors explore how the fusion of terrain information and modelling can bolster national resilience by enabling accurate predictions of various hazards related to climate change, such as heavy rainfall and intensifying typhoons. In this context, smart cities, supported by both city and Earth digital twins, can enhance effective planning of energy production and usage needed for a successful energy transition.

1.3 Keynotes for Further Reading

Japan needs smart energy transition and smart energy solutions. The country which has been playing an important role in the post-war development, in recent years has been hit by different calamities, starting from financial crises to end with severe natural disasters, including the 2011 earthquake and tsunami, struggles with long-know problems like greying society and depopulation. All of them have their energy-related dimensions linked with growing energy prices, post-Fukushima energy mix, or energy poverty.

For Japan, a country with a heavily urbanised population facing these challenges, smart cities offer a promising solution. This agenda aims to support residents, solve problems, improve quality of life, and address various challenges with the help of smart solutions including technology (see Sokołowski and Shimpo, 2025). Among them one may find challenges related to sustainability, climate change, and energy transition. Addressing them in a smart, objective, and systematic manner needs efficient decision-making or long-term planning (see Sethi, Mittal, and Suwa, 2025). This requires processing vast amounts of data, often in real time, across various sectors like electricity, gas, water, waste, or transportation, together with environmental data on GHG, temperature, rainfall, or wildfires, trying to integrate them in open platforms that enable various applications (see Ishii and Hayashi, 2025; Sugita *et al.*, 2025; Yamashita, 2025; Yu *et al.*, 2025). Solutions like advanced smart technologies and AI offer valuable support in this regard (see Yu *et al.*, 2025; Sokołowski and Shimpo, 2025).

These actions must be coordinated among all stakeholders and require a strong sense of collaboration (see Yu *et al.*, 2025). A smart city is a vision created by everyone, regardless of their position within the administrative structure or outside of it – therefore, it is necessary to move beyond the traditional perception of governance (see Piecha, Kanas, and Mikusek, 2025). On the one hand, this may require legal changes that determine the material nature of the city, so its spatial conditions, which are reflected at the administrative level through planning acts (see Zieliński, 2025). On the other hand, policy evaluation needs strengthening to measure sustainability of projects in a clear way with the help of Key Performance Indicators like GHG emission reductions (see Yamashita, 2025).

The essence, as Hiroshi Ito and Kazuhiko Kato (2025) underline, "lies in fostering an improved quality of life within a sustainable framework". Here, smart cities hope "to establish new standards for urban living that prioritise technological innovation and a concerted effort to reduce carbon emissions, promote clean mobility solutions, and support a transition towards a low-carbon economy" (Ito and Kato, 2025). This represents a drive towards sustainability has been a compelling force for many young Japanese, particularly those in urban areas, who have actively advocated for it (Kusanagi, 2025). However, more is needed for the energy transition. A passive stance from the rest of society is insufficient; everyone, not just the youth, must actively demonstrate their commitment to addressing climate change by embracing and promoting smart, sustainable solutions (see Kusanagi, 2025). In this context, education offers positive outcomes, especially when focusing on SDG-related topics in everyday life (see Yu *et al.*, 2025). In a digitalised world, this also necessitates utilising all available channels, including the internet and metaverse. Moreover, to broaden the scope and strengthen the message, sustainable education may require new educators like robots and cybernetic avatars (see Shimpo, 2024; Sokołowski, 2024).

Nevertheless, as Madeline Taylor (2025) writes "achieving Smart City outcomes starts with energy sources". To reduce GHG emissions in Japanese cities, energy production should shift from fossil fuels to electricity generated from non-emission sources like renewables, with grid management helping to balance supply and demand, while residents equipped with photovoltaics and storage batteries can contribute to renewable energy development and help stabilise the grid, with electric vehicles and smart demand-side management solutions further enhancing this transition and allowing households with solar panels to benefit from lower energy bills by consuming the electricity they generate on their rooftops (Kurokawa, 2025).

However, like "there is no single silver bullet for success in creating Smart Cities" (Taylor, 2025), the energy transition "requires discussing different scenarios and solutions" (Kusanagi, 2025). In this light, our book aims to offer specific smart solutions for smart cities, focusing on the energy transition in Japan. We sincerely hope that some of these solutions will find fertile ground beyond Japan, contributing positively to global efforts on climate and energy. This, indeed, is the smart power that Japan possesses.

Acknowledgements

This research was supported by the JST Moonshot R&D project (grant number JPMJMS2215).

References

Fukuda, Y. (2008) "Special Address by Yasuo Fukuda, Prime Minister of Japan, on the Occasion of the Annual Meeting of the World Economic Forum", The World and Japan Database. Available at: https://worldjpn.net/documents/texts/exdpm/20080126.S1E.html (Accessed: 31 August 2024).

Georgescu, M. *et al.* (2021) "Focus on Sustainable Cities: Urban Solutions toward Desired Outcomes", *Environmental Research Letters*, 16(12), p. 120201.

Ishii, H. and Hayashi, Y. (2025) "5Grids Concept as a Basis of Zero-Carbon, Decentralised and Digitalised Society", in M.M. Sokołowski and F. Shimpo (eds) *The Energy Transition in Japan: Smart Cities and Smart Solutions*. London: Routledge, pp. 49–62.

Ito, H. and Kato, K. (2025) "Can Toyota's Woven City Be Considered a Smart City?", in M.M. Sokołowski and F. Shimpo (eds) *The Energy Transition in Japan: Smart Cities and Smart Solutions*. London: Routledge, pp. 187–205.

Kurokawa, S. (2025) "Low-Carbon Transition and Energy Poverty in a Smart City", in M.M. Sokołowski and F. Shimpo (eds) *The Energy Transition in Japan: Smart Cities and Smart Solutions*. London: Routledge, pp. 170–186.

Kusanagi, S. (2025) "E-Methane and Its Future in Japan: City Gas and Smart Cities", in M.M. Sokołowski and F. Shimpo (eds) *The Energy Transition in Japan: Smart Cities and Smart Solutions*. London: Routledge, pp. 81–96.

Piecha, J., Kanas, J.K. and Mikusek, P. (2025) "Central and Local Government Participation in Decentralised Smart City Development Tools as a Manifestation of Modern Methods of Administrative Tasks", in M.M. Sokołowski and F. Shimpo (eds) *The Energy Transition in Japan: Smart Cities and Smart Solutions*. London: Routledge, pp. 120–136.

Sethi, M., Mittal, S. and Suwa, A. (2025) "Developing a Smart Tool for Integrated Climate Action Planning (ICLAP 2050) in Asia-Pacific Cities and Its Application to Japanese Cities", in M.M. Sokołowski and F. Shimpo (eds) *The Energy Transition in Japan: Smart Cities and Smart Solutions*. London: Routledge, pp. 22–48.

Shimpo, F. (2024) "What Are E3LSI Issues in Cybernetic Avatars", in *Ethical, Legal and Social Issues for Symbiotic Society with AI and Robots: Proceedings of the Ninth*

International Conference on Robot Ethics and Standards. 9th ICRES, 29–31 July 2024, Yokohama: CLAWAR, pp. 134–137.

Sokołowski, M.M. (2022) "Artificial Intelligence and Climate-Energy Policies of the EU and Japan", in D. Bielicki (ed.) *Regulating Artificial Intelligence in Industry*. London: Routledge, pp. 138–155.

Sokołowski, M.M. (2024) "Cybernetic Avatars, Robots, and SustAInability", in *Ethical, Legal and Social Issues for Symbiotic Society with AI and Robots: Proceedings of the Ninth International Conference on Robot Ethics and Standards*. 9th ICRES, 29–31 July 2024, Yokohama: CLAWAR, pp. 138–144.

Sokołowski, M.M. and Shimpo, F. (2025) "Japan's Smart Energy Transition: Visions and Solutions for Technology-Driven Efficient Energy Transformation", in M.M. Sokołowski and F. Shimpo (eds) *The Energy Transition in Japan: Smart Cities and Smart Solutions*. London: Routledge, pp. 8–21.

Sokołowski, M.M. and Visvizi, A. (eds) (2023) *Routledge Handbook of Energy Communities and Smart Cities*. London: Routledge.

Sugita, N. *et al.* (2025) "Earth Observation for Digital Twin: Fusion of Smart City and Satellite Data for Japan's Energy Transition", in M.M. Sokołowski and F. Shimpo (eds) *The Energy Transition in Japan: Smart Cities and Smart Solutions*. London: Routledge, pp. 206–230.

Taylor, M. (2025) "Australia Powering Japan's Energy Transition", in M.M. Sokołowski and F. Shimpo (eds) *The Energy Transition in Japan: Smart Cities and Smart Solutions*. London: Routledge, pp. 154–169.

Visvizi, A. and Troisi, O. (eds) (2022) *Managing Smart Cities: Sustainability and Resilience Through Effective Management*. Cham: Springer.

Yamashita, J. (2025) "Could the Mobility as a Service (MaaS) Contribute to the Achievement of Japan's Net-Zero Goal in the Transport Sector?", in M.M. Sokołowski and F. Shimpo (eds), *The Energy Transition in Japan: Smart Cities and Smart Solutions*. London: Routledge, pp. 97–119.

Yu, J. *et al.* (2025) "Sustainable Waste Management for Carbon-Neutral and Circular Economy in Japan", in M.M. Sokołowski and F. Shimpo (eds) *The Energy Transition in Japan: Smart Cities and Smart Solutions*. London: Routledge, pp. 63–80.

Zieliński, P. (2025) "Smart Cities as a Sustainable Development Tool in Spatial Planning Acts of the European Union and Japan: Comparative Analysis", in M.M. Sokołowski and F. Shimpo (eds) *The Energy Transition in Japan: Smart Cities and Smart Solutions*. London: Routledge, pp. 137–153.

2 Japan's Smart Energy Transition

Visions and Solutions for Technology-Driven Efficient Energy Transformation

Maciej M. Sokołowski and Fumio Shimpo

2.1 Introduction

Smart cities can take various forms. However, as underlined in the Japanese *Smart City Guidebook*, in general, they provide "services to support each one of residents using new technologies" and enhance "management in various fields (e.g. planning, development, management/operation)", while also solving "challenges faced by cities and regions ...to create new value" and "[b]eing a sustainable city/region where Society 5.0 is realized ahead of the others" (Cabinet Office, 2021, p. 9).[1] The challenges we face are increasing and have intensified problems such as natural disasters and the disadvantages of growing urbanisation, including traffic congestion, air pollution, water and power supply shortages, as well as sewage and waste disposal issues (see Government of Japan, 2020, p. 2).

Many of these problems are climate-driven or energy-related. Both can be addressed by technology, and cities are a perfect environment for introducing innovations. However, this may require a framework or vision driven by policy goals. Then, programmes and initiatives are needed to develop technologies and innovations, with industry eager to join them. If done correctly, this could represent a successful model for public–private technology cooperation (see Stiglitz and Wallsten, 1999; Liu *et al.*, 2021). To some extent, such fruitful collaboration has been realised in Japan, where different cabinets have managed to work with industry and science to develop Japanese technologies used for the needs of energy, environment, and climate.

In this light, this chapter presents the Japanese policy goals related to climate together with various incentives proposed by Japan at different international forums by policy leaders – Prime Ministers Kishida, Takeshita, Miyazawa, Hashimoto, Abe, and Fukuda.[2] As examined in our chapter, many of these proposals have been focused on technology, aimed at developing innovative solutions to solve global environmental and climate problem. In this chapter we link this approach with *monozukuri* – a profound commitment to manufacturing excellence – that we discuss with respect to energy transition in Japan.

In this context, we highlight public-driven programmes that stimulated Japanese research and development (R&D) capacity in the energy sector (the Sunshine

DOI: 10.4324/9781003471417-2

Project, the Moonlight Project, and the New Sunshine Project), as well as present other policy incentives focused on technological innovations developed to address climate change, such as Green IT. Moreover, in this chapter we offer a closer look at energy efficiency as a key component of the Japanese technology-oriented approach. In conclusion, we combine these two elements (energy efficiency and the focus on technology) and analyse them in the context of Japan's energy transition with smart cities at its heart.

2.2 Aims of Japanese Climate Policy: Visions of Japanese Political Leaders Since 1980s

> Japan aims to reduce its greenhouse gas emissions by 46% by 2030, and will continue strenuous efforts in its challenge to meet the lofty goal of cutting its emission by 50%. Japan has already achieved a reduction of approximately 20%, Japan's reduction is on track.

In these words, Prime Minister Fumio Kishida (2023) addressed the aims of Japanese climate action at the Climate Conference COP28 in Dubai. He concluded his speech with "Japan will dedicate itself to undertaking actions on climate change which is our common agenda in cooperation with other countries and leading the efforts of the international community".

Fumio Kishida is not the first head of the Japanese government who has found climate action important. Back in the late 1980s, Prime Minister Noboru Takeshita became aware of climate change as a significant issue in international relations (see Kameyama, 2010, p. 192). He placed global environmental problems on the political agenda of the Liberal Democratic Party (LDP) (see Okano-Heijmans, 2012, pp. 341–342). This stance was maintained by subsequent governments, although the Japanese position on climate often varied from that of global partners, particularly in terms of commitment to emissions reduction targets (see Kawashima, 2001, pp. 169–170). The difference, generally consisting of Japan's preference for a more light-handed approach, has been the subject of criticism both at home and abroad (see Sokołowski, 2022b, pp. 90–92).[3]

These circumstances led to a general consensus among most of the Japanese political leaders and business stakeholders that they should take a more active role in tackling climate-related problems (see Żakowski, Bochorodycz and Socha, 2018, p. 183). As a result, the Action Programme to Arrest Global Warming (Government of Japan, 1990) was adopted as the baseline Japanese position in terms of participating in the creation of an international framework aimed at reducing carbon dioxide (CO_2) emissions to 1990 levels by the year 2000 (see Fermann, 1993, p. 292; Shindo, Hakuta and Miyama, 1992, pp. 773–775). In this context, apart from a reference to the Action Programme, in a written statement of Prime Minister Kiichi Miyazawa distributed during the United Nations Conference on Environment and Development (UNCED) held in Rio de Janeiro, commonly known as the Earth Summit,[4] one will find following words:

> [t]he world is now at a major turning point. We are now searching for a new international order which values the well-being of each and every person; an order in which human dignity is fully respected by upholding the principles of freedom, democracy and sustainable development. We should aim at constructing a new era in which we all live as global citizens.
>
> It is the most fundamental prerequisite of this "era of global citizenry" that environmental protection and sustainable development be achieved in tandem.
>
> (Miyazawa, 1992)

The 1992 summit resulted in establishing the cornerstone of such a "new international order" – the United Nations Framework Convention on Climate Change (1992). In 1995, the parties to the Framework Convention met in Berlin during the 1st Conference of Parties (COP) to discuss their commitments related to climate change (see Oberthür and Ott, 1995). Two years later, in Kyoto, delegates gathered at COP3 concluded the complex multilateral negotiations by adopting the Kyoto Protocol to the United Nations Framework Convention on Climate Change (1997) with its legally binding targets for greenhouse gas emission reductions (see Yamin, 1998). Just a few months earlier, at the Special Session of the United Nations General Assembly, Prime Minister Ryutaro Hashimoto appealed for united global action to address climate change – a pressing challenge that directly impacted people's lives:

> [i]t goes without saying that we must strive, also from a medium- and long-term perspective, to solve the issue of global climate change. For example, if we want to stabilize the density of carbon dioxide in the atmosphere at a level about twice as high as it was before the Industrial Revolution, it will be necessary to reduce global per capita carbon dioxide emissions to one ton by the year 2100. This is a great challenge that cannot be met with the existing technology.
>
> (Hashimoto, 1997a)

To accelerate efforts to combat global warming, Prime Minister Hashimoto presented the Green Initiative, named the Comprehensive Strategy for the Prevention of Global Warming. Envisioned as an international effort, this initiative rested on two pillars: Green Technology and Green Aid. Also in his speech at COP3, Prime Minister Hashimoto returned to the "need to accelerate the development of innovative technologies through international cooperation" (Hashimoto, 1997b). This agenda represents the essence of Japan's post-1980s economic diplomacy, focusing on enhanced economic cooperation in green energy and environmental technologies (see Okano-Heijmans, 2012, p. 342). It aimed to promote actions such as the development of energy-efficient and renewable energy technologies, alongside innovations for environmental protection, as well as to provide financial assistance to developing countries to tackle energy problems and global warming (Hashimoto, 1997a).

Another notable policy statement on climate made at the highest level of the Japanese government was Prime Minister Fukuda's address at the 2008 World Economic Forum in Davos. In a very direct manner, he labelled climate changed a "top priority" accompanied by global environmental issues that "have now gone beyond the discussion stage to become real problems with significant effects on our day-to-day lives and economic activities" (Fukuda, 2008). "This constitutes a major new challenge for humanity, as we could be courting catastrophe in both the natural environment and our socio-economic activities if we stand by and do nothing" (Fukuda, 2008). Although said more than 15 years ago, this still sounds very current.

Apart from declarations, during the 2008 summit in Davos, Prime Minister Fukuda presented his Cool Earth Promotion Programme. The 2008 programme was a follow-up of the Japanese 2007 initiative under Prime Minister Abe called the Cool Earth 50 Plan. The 2007 plan proposed reducing global greenhouse gas emissions by half by the middle of the century, while developing innovative technologies and building a low-carbon society (Kikkawa, 2009). This post-Kyoto framework was based on the participation of all major emitters, with their flexibility and diversity, as well as compatibility between environmental protection and economic growth (Kikkawa, 2009).

In the same vein, the proposed 2008 programme was aimed at establishing a post-Kyoto regime on global reduction of emissions. Prime Minister Fukuda (2008) called for a fair mechanism with equitable emission targets in which everyone participates, including all major emitters. For these reasons, he explained:

> [t]he target could be set based on a bottom-up approach by compiling on sectoral basis energy efficiency as a scientific and transparent measurement and tallying up the reduction volume that would be achieved based on the technology to be in use in subsequent years. The base year should also be reviewed from the standpoint of equity. Without equity, it will be impossible to maintain efforts and solidarity over the long term.
>
> (Fukuda, 2008)

Another element of "solidarity" proposed by Prime Minister Fukuda (2008) was the International Environment Cooperation. This incentive involved providing 10 billion USD in aid to developing countries, aiming to achieve emission reductions and foster economic growth (Fukuda, 2008). Japan distributed this aid through a new financial mechanism called the Cool Earth Partnership, which rested on two pillars: first, there was grant aid to support adaptation to climate change and enhance access to clean energy; second, loans were provided for climate change mitigation actions (Ministry of Foreign Affairs, 2008). The partnership covered such actions as technical assistance, feasibility studies, co-benefit projects, or implementing programmes to address global warming in developing countries. On the Japanese side, actors involved in the Cool Earth Partnership included the Japan Bank for International Cooperation (JBIC), the New Energy and

Industrial Technology Development Organization (NEDO), and the Nippon Export and Investment Insurance (NEXI) (Ministry of Foreign Affairs, 2008). Among the beneficiaries of this incentive one could find such countries as Senegal, Guyana, Madagascar, Indonesia, Bangladesh, Mozambique, Niger, Ethiopia, and Kenya (Kiko Network, 2010).

The actions undertaken within the framework of the Cool Earth Partnership followed the 2008 words of Prime Minister Fukuda: "[w]hat Japan can take action is to transfer high quality environmental technology to a greater number of countries". This was correlated with another aspect of his vision: innovative technologies to be applied for a transition to a low-carbon society. According to the then-head of the Japanese government, the "breakthroughs in technological innovation" were "absolutely critical" to cut emissions by half, although this was "a very challenging task" that required "a tremendous investment in technology" (Fukuda, 2008). Prime Minister Fukuda (2008) highlighted some of the proposed innovations – the list included zero-emission coal-fired power plants; low-cost, high-efficiency solar installations; and green information technologies (Green IT). He also announced investments in R&D in the fields of environment and energy, totalling 30 billion USD over the next five years (Fukuda, 2008).

2.3 Innovative Technologies for Energy, Environment, and Climate: Japan's Approach

The idea of a technology-driven transition and finding solutions in innovations has its roots deeply embedded in Japanese history (see Asai, Hayashi and Minazuki, 2011). Over the years, the Japanese approach to environment, climate, and energy transition has been technology-oriented. One may find here some elements of *monozukuri* – "thing-making"[5] that includes a sense of artisanship and manufacturing deriving from the spirit of making things with dedication and commitment, aiming for improvement (Sokołowski, 2022a, p. 139). The authors of a study on environmental performance of the Japanese automotive producer Toyota write about *monozukuri* as "manufacturing that is in harmony with nature and is value adding for the society", referring to *monozukuri* as "the older sister of sustainable manufacturing" (Zokaei, Manikas and Lovins, 2017, p. 381).

In the context of energy transition, this approach could be linked with enhancing energy efficiency of the Japanese assembly industries such as automotive and electronics, as well as developing sustainable solutions transferred abroad to contribute to saving energy and reducing CO_2 emissions on a global scale (see Komiyama, 2014, p. 50). As numerous declarations, policies, and strategies have shown, this model has been promoted at various levels by Japan for a long time. The Japanese energy industry serves as a good example. Since the 1970s, Japan has been supporting this sector with various public-driven programmes like the Sunshine Project, the Moonlight Project, or the New Sunshine Project to stimulate country's R&D capacity in the energy sector (Sokołowski, 2021, p. 150). Focusing

Table 2.1 Leading Japanese R&D incentives for renewable energy and energy efficiency

Name	*Starting year*	*Budget*	*Description*
Sunshine Project	1974	From about 2.4 billion JPY (1974) to 44 billion JPY (1985) with less than 30 billion JPY in final years (1991)	Project initiated by the Agency of Industrial Science and Technology under the Ministry of International Trade and Industry, which has offered R&D activities in different energy activities, including solar – seen as an alternative to conventional fossil fuels – with project's support the solar cell production in Japan reached 12,500 kW in 1986
Moonlight Project	1978	From less than 2 billion JPY (1978) to around 12 billion JPY (1991)	Project was created to develop energy conservation technologies combining R&D works of government and industry
New Sunshine Project	1993	56 billion JPY (1996)	In response to the 1990s stagnation trends and to take advantage of the potential of renewable energy in Japan, the Sunshine Project along with the Moonlight Project and the Global Environmental Technology Program evolved into the New Sunshine Project

Source: Sokołowski and Kurokawa (2024, p. 360).

on solar energy and energy savings, they significantly contributed to advancing renewables and energy efficiency in Japan (see Table 2.1).

An important role was played by the Ministry of International Trade and Industry (MITI) and its branches like the Agency of Industrial Science and Technology. This was just one component of a larger framework with MITI at its core – "a sophisticated policy system" that fostered to technological development of Japan, starting a chain reaction that ultimately boosted the entire Japanese industry and contributed to Japan's economic growth (Watanabe and Honda, 1992, pp. 47–48). The framework derived from the 1971 vision for the 1970s, projected by MITI to be the step for a mature economy sector (Committee on the History of Japan's Trade and Industry Policy RIETI, 2020, p. 21). In MITI's complex policy system, the role of visions was to stimulate R&D by perceiving directions, identifying long-term goals, as well as creating consensus, instilling confidence, and establishing shared responsibilities among the broad sectors concerned (Wakabayashi, Griffy-Brown and Watanabe, 1999, p. 4). The visions were not limited to anticipating future possibilities or preferences but rather aimed to synchronise the "expected futures, possible futures or preferred futures" (Wakabayashi, Griffy-Brown and Watanabe, 1999, p. 4).

The vision for the 1970s emphasised a significant role of innovative R&D aimed at reducing Japan's dependency on materials and energy as well as highlighted the importance of energy conservation and recycling of resources in a long-term perspective (Watanabe, 1995, p. 453). In practice, adding energy conservation to the new agenda reoriented Japan's transition from traditional heavy manufacturing industry , leading to a greater focus on machine sector (RIETI, 2020, p. 21). In this way, the vision for the 1970s was a new approach that proposed a transition of Japanese economy towards a knowledge-based industrial structure aimed at reducing the environmental burdens and relying less on energy and materials and more on technology (Watanabe, 1995, p. 453). This contrasted with post-war development driven by highly material- and energy-intensive industries (see Watanabe, 1995, pp. 452–453). The rapid economic growth came at a significant cost: pollution. During the 1950s, Japanese industrial cities experienced extremely high levels of dust coming from Japan's main post-war energy source – coal, and when, in the 1960s, coal was gradually replaced by oil, sulphur oxides emissions emerged as another environmental concern (Imura, 2005, pp. 21–22).

Having impact on people's health, as exemplified by the "four big pollution diseases"[6] – Itai-itai disease, Minamata disease, Niigata Minamata disease, and Yokkaichi asthma – the unlimited industrial growth became a visible problem (see Sokołowski, 2022b, pp. 86–87). In response to the growing public concern about environmental and health issues, which put pressure on government and politicians, a series of new pollution-related laws were adopted in the late 1960s and early 1970s (see Sokołowski and Kurokawa, 2022, p. 185). Many of these laws were passed during a memorable 1970 session of the Japanese parliament, which has since become known as the Pollution Diet (see Nakanishi, 2016, p. 2; Sokołowski, 2022b, p. 88). One of the legislative changes introduced during the special session was the removal of the controversial economic harmony clause, which mandated a balance between pollution control and economic development (Tsuji, 2010, p. 345; Nakanishi, 2016, p. 2).

The environmental concerns were also among reasons of the 1970s shift in the Japanese industrial policy (see Watanabe, 2002, p. 236). To identify the basic policy elements for implementing the vision for 1970s, in 1971 MITI established an ecology research group (Watanabe, 2002, pp. 236–237). As a result of its research work, this group proposed a comprehensive method for analysing and evaluating the complex mutual relations between humans, industry, and the environment (Watanabe, 1995, p. 453). In 1973, MITI outlined its new approach with efforts directed to R&D programmes aimed at creating an environmentally friendly and efficient energy system (Watanabe, 1995, p. 453, 2002, p. 237).

In this context, in 1974, the Sunshine Project, the first in a series of initiatives focused on new energy technologies, was introduced (Sokołowski, 2022b, p. 142). Aiming to develop clean energy technologies by the year 2000, the 1974 project anticipated increasing the Japan's energy supply from new energy resources to around 5% by 1990 (Kamimoto and Hayashi, 1982, p. 186; Matsumoto, 2005, p. 624). Regardless of its name, the Sunshine Project supported basic research by providing funding for conceptual design, component technology development, and the construction and operation of pilot and, eventually, commercial plants in the fields of solar energy, coal gasification and liquefaction, hydrogen energy, as well as ocean and wind power (see Kamimoto and Hayashi, 1982, p. 186; Matsumoto, 2005, p. 625; Sokołowski, 2022b, p. 142).

Four years later, in 1978, Japan launched another energy-focused initiative, the Moonlight Project, this time centred on energy conservation (see Noguchi, 1994, p. 1342). The 1978 project aimed to conduct joint government–industry R&D on energy-saving technologies that were too costly or risky for private companies to implement solely with the use of own capital (see Fukasaku, 1995; Sokołowski, 2022b, pp. 191–192). This included technologies like high-performance gas turbines, heat pumps, fuel cells, or batteries (see Noguchi, 1994, p. 1342). The aim was to achieve higher efficiency, reduce environmental pollution, and promote the use of versatile fuels (see Hattori, Masuda and Ozawa, 1993). Among companies that participated in this project, one can list Fuji Electric Corporate Research and Development, Sanyo Electric, Mitsui Shipping, Fujikura Corporation, Mitsubishi Heavy Industries, or Murata Manufacturing (see Tagawa, 1993, pp. 7–9).

The magnitude of energy efficiency's potential is demonstrated by the remarks made by the Prime Minister Fukuda, almost 30 years after the introduction of the Moonlight Project:

> [i]t goes without saying that aiming at the most efficient use of energy is now an obligation upon humanity. For the time being, until when innovative technologies that will drastically reduce the emissions of greenhouse gas become practically available, the whole world must make efforts to maximize the improvement of energy efficiency.
>
> … Japan has been committed as a nation to energy conservation. Over the last 30 years, we have succeeded in doubling our real GDP without increasing the overall energy consumption of the industry sector. This demonstrates that Japan succeeded in the simultaneous pursuit of both economic growth and environmental protection.
>
> (Fukuda, 2008)

This success was substantiated by data. As the International Energy Agency (IEA) reported in its Japan 2008 review, between 1973 and 2005, Japanese energy efficiency, measured in terms of energy consumption per unit of production, improved by 20% in steel industry, 52% in paper-making industry, 24% in cement industry, and 29% in chemical industry (IEA, 2008a, p. 54). This placed Japan's major industrial sectors among the most energy efficient within the IEA countries (IEA,

2008a, p. 54).[7] Japan's technology-oriented policy has garnered global recognition. Former executive director of the IEA, Nobuo Tanaka, summarised this as follows:

> Japan deserves significant praise for its technology leadership – which is helping drive enhanced energy security and climate change mitigation throughout the globe. The government continues to raise the profile of energy technology, including international collaboration and transfer, and has placed energy, climate change and sustainability at the top of its agenda for the G8 meetings it is hosting.
>
> (IEA, 2008b)

As Prime Minister Fukuda highlighted in 2008, the technological innovations developed to address climate change had an IT dimension, often referred to as Green IT (see Fukuda, 2008). This global trend, which emerged in the 2000s, was a response to diverse environmental issues associated with IT like e-waste and the widespread use of various chemicals in the industry, but its primary emphasis was addressing the climate impact of the IT sector (Fors, Kreps and O'Brien, 2024, p. 33). It found its way to Japan around the turn of 2007 and 2008, when first, in 2007, the Ministry of Economy, Trade and Industry (METI) established the Green IT Initiative, and then, in 2008, METI founded the Green IT Promotion Council (Majima *et al.*, 2017, pp. 82–83). The council primarily focused on energy conservation by promoting the use of IT energy-saving products and technologies with high energy efficiency while evaluating the sector in this regard (see Green IT Promotion Council, 2009). In this context, Majima *et al.* (2017, pp. 91–93) argue that in Japan, Green IT was a by-product of a stronger actor-network focused on energy saving that eventually paved the way for other concepts such as smart-related concepts like smart grid, smart city, or smart community. These smart concepts where electricity is used intelligently, gained prominence in Japan following the devastating earthquake and tsunami of March 2011, eclipsing Green IT initiatives (see Majima *et al.*, 2017, p. 91).[8]

Like the Japanese climate policy and approach to innovations, the idea of smart cities is closely tied to energy-saving and enhancing energy efficiency – using energy in a smart way. In this context, the crucial policy element upon which smart cities are to be built in Japan comprises smart communities. As outlined in the 4th Strategic Energy Plan, creating smart communities can significantly improve energy supply efficiency through demand response systems, leading to substantial energy savings (see METI, 2014, p. 74). As stated in the Long-term Strategy under the Paris Agreement, "Japan is developing smart cities that adopt the energy management system made possible by the digital revolution" (Government of Japan, 2019, p. 61).

Using digital technologies for energy transition is also a key element of smart city, defined in Japan as both sustainable and progressing towards the realisation of Society 5.0 (see Cabinet Office, 2021, p. 9). In this vision, the relationship between science and technology will have a major impact on driving innovation and development, with artificial intelligence (AI) playing a particularly important role (see Shimpo, 2018, p. 49). Moreover, the application of AI in energy transition demonstrates the close relationship between green transformation (GX) and digital transformation (DX) in Japan (see Sokołowski, 2022a). This also highlights the interconnections of Society 5.0, GX, DX, and smart cities, where "[s]mart cities are at the heart of the digital economy and will be crucial for the realization of a green economy", as D. Hugh Whittaker (2024, p. 83) underlines.

2.4 Conclusion

Over the years, Japanese energy and climate policies have been shaped by the visions of different leaders and administrations. While these agendas have varied in detail due to the differing circumstances in which they have been proposed, they share some clear similarities. A focus on technologies and energy efficiency has been a consistent element of these agendas, driving the realisation of energy and climate policies in a way that closely aligns with the assumptions of smart cities. Saving energy, whether through traditional methods or modern efficiency techniques, has long been a key component of Japan's approach to addressing energy and climate challenges. This "smart use of energy" emphasises conservation and efficiency that have become the pillars of smart-related initiatives in Japan, including smart cities and smart communities.

Finally, the Japanese approach offers another crucial lesson for the ongoing AI revolution. We must urgently address sustainability issues associated with AI, such as energy-intensive AI systems that inefficiently consume an excessive amount of power. *Smart cities* that rely on these types of AI solutions are, in fact, *not smart*. We must prioritise energy efficiency as a fundamental aspect of a successful transition. This concerns both GX and DX. Solutions for addressing this issue vary widely, ranging from legal frameworks with different degrees of impact to softer measures such as education.[9] Here too, we have to choose smart.

Acknowledgments

This research was supported by the JST Moonshot R&D project (grant number JPMJMS2215).

Notes

1 Society 5.0 (the super smart society) is the new society created by transformations led by scientific and technological innovation, after hunter–gatherer society – Society 1.0; agricultural society – Society 2.0; industrial society – Society 3.0; and information society – Society 4.0 (see Shimpo, 2018, p. 49).

2 The sequence as discussed in the chapter.

3 Japanese political leaders tend to navigate between two influential voices: one resonating from domestic stakeholders and citizens through elections, and the other echoing from beyond Japan's borders, also called *gaiatsu* (in Japanese: 外圧) – an external pressure (see Kameyama, 2010, p. 192).

4 The "Earth Summit" took place in Rio de Janeiro, Brazil, in June 1992. However, Prime Minister Miyazawa did not attend the event (see Kameyama, 2010, p. 192); instead, Shozaburo Nakamura, who served as the Director-General of the Environmental Agency at the time, represented the Japanese government – the prime minister's speech was distributed at the venue, ensuring an official record (Ministry of the Environment, 1993).

5 In Japanese: ものづくり.

6 In Japanese: 四大公害病 [yondaikōgaibyō].

7 As of the publication date of the 2008 IEA report, the following countries were its members: Australia, Austria, Belgium, Canada, Czech Republic, Denmark, Finland, France, Germany, Greece, Hungary, Ireland, Italy, Japan, Republic of Korea, Luxembourg, the Netherlands, New Zealand, Norway, Portugal, Slovak Republic, Spain, Sweden, Switzerland, Turkey, the United Kingdom, and the United States.

8 Although these concepts assumed greater importance in Japan after March 2011, they were already being applied in Japan before the Great East Japan Earthquake, as highlighted in the Japanese Basic Energy Plans/Strategic Energy Plans (see METI, 2010, p. 49, 2014, pp. 74–75). It is worth noting that since the fourth plan, METI and its Agency for Natural Resources and Energy have been using the English translation "Strategic Energy Plans" (see Agency for Natural Resources and Energy, 2024).

9 In the context of education, AI can provide valuable solutions, such as cybernetic avatars serving as sustainability agents or climate educators. This is particularly important in the context of a cyberspace with avatars as its elements (see Sokołowski, 2024), or, in other words, creating *digital smart cities*. Frameworks addressing a wide range of problems, such as ethical, economic, environmental, legal, and social implications (E^3LSI), offer some promising solutions (see Shimpo, 2024).

References

Agency for Natural Resources and Energy (2024) "Strategic Energy Plan". Available at: www.enecho.meti.go.jp/en/category/others/basic_plan/ (Accessed: 31 July 2024).

Asai, T., Hayashi, H. and Minazuki, A. (2011) "Fostering a 'Monozukuri (Manufacturing)' Organization Suitable for the 21st Century Digital Economy", in P. Parucha (ed.) *New Knowledge in a New Era of Globalization*. Rijeka: InTech, pp. 199–224.

Cabinet Office (2021) "Smart City Guidebook". Available at: www8.cao.go.jp/cstp/society5_0/smartcity/01_scguide_eng_1.pdf (Accessed: 22 July 2024).

Committee on the History of Japan's Trade and Industry Policy RIETI (2020) *Dynamics of Japan's Trade and Industrial Policy in the Post Rapid Growth Era (1980–2000)*. Singapore: Springer (Economics, Law, and Institutions in Asia Pacific).

Fermann, G. (1993) "Japan's 1990 Climate Policy Under Pressure", *Security Dialogue*, 24(3), pp. 287–300.

Fors, P., Kreps, D. and O'Brien, A. (2024) "Green IT: The Evolution of Environmental Concerns Within ICT Policy, Research and Practice", in T. Lynn *et al.* (eds) *Digital Sustainability*. Cham: Springer, pp. 25–48.

Fukasaku, Y. (1995) "Energy and Environment Policy Integration: The Case of Energy Conservation Policies and Technologies in Japan", *Energy Policy*, 23(12), pp. 1063–1076.

Fukuda, Y. (2008) "Special Address by Yasuo Fukuda, Prime Minister of Japan, on the Occasion of the Annual Meeting of the World Economic Forum", The World and Japan Database. Available at: https://worldjpn.net/documents/texts/exdpm/20080126.S1E.html (Accessed: 22 July 2024).

Government of Japan (1990) "Action Programme to Arrest Global Warming". In Japanese: 地球温暖化防止行動計画 [chikyūondankabōshi kōdō keikaku].

Government of Japan (2019) "The Long-Term Strategy Under the Paris Agreement".

Government of Japan (2020) "Japan's Smart Cities: Solving Global Issues Such as the SDGs, etc. Through Japan's Society 5.0". Available at: www.kantei.go.jp/jp/singi/keikyou/pdf/Japan's_Smart_Cities-1(Main_Report).pdf.

Green IT Promotion Council (2009) "Activities – Green IT Promotion Council". Available at: https://home.jeita.or.jp/greenit-pc/e/about/activity.html (Accessed: 22 July 2024).

Hashimoto, R. (1997a) "Statement by Prime Minister of Japan Ryutaro Hashimoto at UNGASS", The World and Japan Database. Available at: https://worldjpn.net/documents/texts/exdpm/19970623.S1E.html (Accessed: 22 July 2024).

Hashimoto, R. (1997b) "The Opening Statement by Prime Minister Ryutaro Hashimoto at the High-Level Segment Attended by Ministers and Other Heads of Delegation of the Third Session of the Conference of the Parties to the United Nations Framework Convention on Climate Change", Ministry of Foreign Affairs of Japan. Available at: www.mofa.go.jp/policy/environment/warm/cop3/hashimoto.html (Accessed: 22 July 2024).

Hattori, M., Masuda, M. and Ozawa, T. (1993) "Development of Ceramic Gas Turbine Components for CGT301 in Moonlight Project", in *Proceedings of the ASME 1993 International Gas Turbine and Aeroengine Congress and Exposition. Volume 1: Aircraft Engine; Marine; Turbomachinery; Microturbines and Small Turbomachinery. The International Gas Turbine and Aeroengine Congress and Exposition*, 24–27 May 1993, Cincinnati, OH: ASME, p. V001T04A005.

IEA (2008a) "Energy Policies of IEA Countries – Japan 2008 Review". Paris: IEA. Available at: https://iea.blob.core.windows.net/assets/124781b2-f54c-4e8c-af32-4f389eb592d4/EnergyPoliciesofIEACountriesJapan2008.pdf (Accessed: 22 July 2024).

IEA (2008b) "The IEA Praises Japan's Commitment to and Action on Technology Development and Deployment Across the Globe, and Urges the Country to Strengthen Domestic Energy and Climate Policies". Paris: IEA. Available at: www.iea.org/news/the-iea-praises-japans-commitment-to-and-action-on-technology-development-and-deployment-across-the-globe-and-urges-the-country-to-strengthen-domestic-energy-and-climate-policies (Accessed: 22 July 2024).

Imura, H. (2005) "Japan's Environmental Policy: Past and Future", in H. Imura and M. Schreurs (eds) *Environmental Policy in Japan*. Cheltenham: Edward Elgar Publishing, pp. 15–48.

Kameyama, Y. (2010) "Japan in the Midst of Multilateral Negotiations on the Future Framework for Climate Change", in *Comparative Research on Major Regional Powers in Eurasia-Paper 2*. Hokkaido Slavic-Eurasian Research Center, pp. 187–199. Available at: www.ceeol.com/search/chapter-detail?id=561694 (Accessed: 3 September 2024).

Kamimoto, M. and Hayashi, H. (1982) "Sunshine Project Solar Photovoltaic Program and Recent Activities in Japan", *International Journal of Solar Energy*, 1(3), pp. 185–195.

Kawashima, Y. (2001) "Japan and Climate Change: Responses and Explanations", *Energy & Environment*, 12(2/3), pp. 167–179.

Kikkawa, T. (2009) *Japan's Contribution to Cool Earth*. Working Paper Series 99. Tokyo: Center for Japanese Business Studies, Graduate School of Commerce and Management, Hitotsubashi University.

Kiko Network (2010) "The Situation of the Cool Earth Partnership and the Hatoyama Initiative". Available at: www.kikonet.org/kiko/wp-content/uploads/2010/05/20100524_CEP_and_HIEng.pdf (Accessed: 22 July 2024).

Kishida, F. (2023) "Statement by Prime Minister Kishida Fumio at COP28 World Climate Action Summit", Prime Minister's Office of Japan. Available at: https://japan.kantei.go.jp/101_kishida/statement/202312/01statement.html (Accessed: 22 July 2024).

Komiyama, H. (2014) *Beyond the Limits to Growth: New Ideas for Sustainability from Japan*. Tokyo: Springer.

Kyoto Protocol to the United Nations Framework Convention on Climate Change (1997).

Liu, T. *et al.* (2021) "Emerging Themes of Public-Private Partnership Application in Developing Smart City Projects: A Conceptual Framework', *Built Environment Project and Asset Management*, 11(1), pp. 138–156.

Majima, T. *et al.* (2017) "Green IT Did Not Take Place: The Translation of Environmentally Friendly IT in Japan", *Journal of Information and Management*, 37(2), pp. 81–96.

Matsumoto, M. (2005) "The Uncertain but Crucial Relationship Between a 'New energy' Technology and Global Environmental Problems: The Complex Case of the 'Sunshine' project", *Social Studies of Science*, 35(4), pp. 623–651.

METI (2010) "Basic Energy Plan". In Japanese: エネルギー基本計画 [enerugī kihon keikaku].

METI (2014) "Strategic Energy Plan".

Ministry of Foreign Affairs (2008) *MOFA: Financial Mechanism for "Cool Earth Partnership"*. Available at: www.mofa.go.jp/policy/economy/wef/2008/mechanism.html (Accessed: 22 July 2024).

Ministry of the Environment (1993) "Section 2. 1. Outcomes of the Earth Summit: 1993 Environmental White Paper". In Japanese: 第2節.1 地球サミットの成果: 平成5年版環境白書 [dai ni setsu ichi chikyū samitto no seika heisei go-nen-ban kankyō hakusho]. Available at: www.env.go.jp/policy/hakusyo/h05/9368.html (Accessed: 22 July 2024).

Miyazawa, K. (1992) "Speech by Prime Minister Kiichi Miyazawa at the United Nations Conference on Environment and Development (UNCED)", The World and Japan Database. Available at: https://worldjpn.net/documents/texts/exdpm/19920613.S1E.html (Accessed: 22 July 2024).

Nakanishi, Y. (2016) "Introduction: The Impact of the International and European Union Environmental Law on Japanese Basic Environmental Law", in Y. Nakanishi (ed) *Contemporary Issues in Environmental Law: The EU and Japan*. Tokyo: Springer, pp. 1–15.

Noguchi, T. (1994) "The Status of Solar Energy Systems in Japan", *Renewable Energy*, 5(5–8), pp. 1342–1349.

Oberthür, S. and Ott, H. (1995) "The First Conference of the Parties", *Environmental Policy and Law*, 25(4–5), pp. 144–156.

Okano-Heijmans, M. (2012) "Japan's 'Green' Economic Diplomacy: Environmental and Energy Technology and Foreign Relations", *The Pacific Review*, 25(3), pp. 339–364.

Shimpo, F. (2018) "The Principal Japanese Ai and Robot Law, Strategy and Research Toward Establishing Basic Principles", *Journal of Law and Information System*, 3, pp. 44–65.

Shimpo, F. (2024) "What Are E3LSI Issues in Cybernetic Avatars", in *Ethical, Legal and Social Issues for Symbiotic Society with AI and Robots: Proceedings of the Ninth International Conference on Robot Ethics and Standards. 9th ICRES*, 29–31 July 2024, Yokohama: CLAWAR, pp. 134–137.

Shindo, Y., Hakuta, T. and Miyama, H. (1992) "The Present Status of Carbon Dioxide Removal in Japan", *Energy Conversion and Management*, 33(5), pp. 773–779.

Sokołowski, M.M. (2021) "Models of Energy Communities in Japan (Enekomi): Regulatory Solutions from the European Union (Rescoms and Citencoms)", *European Energy and Environmental Law Review*, 30(4), pp. 149–159.

Sokołowski, M.M. (2022a) "Artificial Intelligence and Climate-Energy Policies of the EU and Japan', in D. Bielicki (ed.) *Regulating Artificial Intelligence in Industry.* London: Routledge, pp. 138–155.

Sokołowski, M.M. (2022b) *Energy Transition of the Electricity Sectors in the European Union and Japan: Regulatory Models and Legislative Solutions.* Cham: Palgrave Macmillan.

Sokołowski, M.M. (2024) "Cybernetic Avatars, Robots, and SustAInability", in *Ethical, Legal and Social Issues for Symbiotic Society with AI and Robots: Proceedings of the Ninth International Conference on Robot Ethics and Standards. 9th ICRES*, 29–31 July 2024, Yokohama: CLAWAR, pp. 138–144.

Sokołowski, M.M. and Kurokawa, S. (2022) "Energy Justice in Japan's Energy Transition: pillars of just 2050 carbon neutrality", *Journal of World Energy Law & Business*, 15, pp. 183–192.

Sokołowski, M.M. and Kurokawa, S. (2024) "The Bamboo That Bends Is Stronger Than the Oak That Resists: Active Energy Consumers in Japan", in T. Soliman Hunter *et al.* (eds) *Routledge Handbook of Consumer Protection and Behaviour in Energy Markets.* London: Routledge, pp. 356–367.

Stiglitz, J.E. and Wallsten, S.J. (1999) "Public–Private Technology Partnerships: Promises and Pitfalls", *American Behavioral Scientist*, 43(1), pp. 52–73.

Tagawa, H. (1993) "Status of SOFC Development in Japan", *Proceedings of the Electrochemical Society*, 1993(1993–4), pp. 6–15.

Tsuji, Y. (2010) "The Legal Issues on Environmental Administrative Lawsuits Under the Amendment of ACLA in Japan", *Yonsei Law Journal*, 1(2), pp. 339–362.

United Nations Framework Convention on Climate Change, adopted 9 May 1992, entered into force 21 March 1994 (1992).

Wakabayashi, K., Griffy-Brown, C. and Watanabe, C. (1999) "Stimulating R&D: An Analysis of the Ministry of International Trade and Industry's 'Visions' and the Current Challenges Facing Japan's Technology Policy-Making Mechanisms", *Science and Public Policy*, 26(1), pp. 2–16.

Watanabe, C. (1995) "Mitigating Global Warming by Substituting Technology for Energy: MITI's Efforts and New Approach", *Energy Policy*, 23(4–5), pp. 447–461.

Watanabe, C. (2002) "Industrial Ecology and Technology Policy: Japanese Experience", in R.U. Ayres and L.W. Ayres (eds) *A Handbook of Industrial Ecology*. Cheltenham: Edward Elgar, pp. 232–245.

Watanabe, C. and Honda, Y. (1992) "Japanese Industrial Science & Technology Policy in the 1990s: MITI's Role at a Turning Point", *Japan and the World Economy*, 4(1), pp. 47–67.

Whittaker, D.H. (2024) *Building a New Economy: Japan's Digital and Green Transformation*. Oxford: Oxford University Press.

Yamin, F. (1998) "The Kyoto Protocol: Origins, Assessment and Future Challenges", *Review of European Community & International Environmental Law*, 7(2), pp. 113–127.

Żakowski, K., Bochorodycz, B. and Socha, M. (2018) *Japan's Foreign Policy Making: Central Government Reforms, Decision-Making Processes, and Diplomacy*. Cham: Springer.

Zokaei, K., Manikas, I. and Lovins, H. (2017) "Environment is Free; But It's Not a Gift", *International Journal of Lean Six Sigma*, 8(3), pp. 377–386.

3 Developing a Smart Tool for Integrated Climate Action Planning (ICLAP 2050) in Asia-Pacific Cities and Its Application to Japanese Cities

Mahendra Sethi, Shilpi Mittal, and Aki Suwa

3.1 Introduction

Since the mid-2000s, smart cities have been considered as a new urban paradigm gaining popularity in policy and business circles, finding equal traction amongst government and private sector. The academia too has been gradually focusing on smart cities as an emerging area of research and specialised courses (CIDOB, 2014). It is widely seen that smartness is involved in data collection, often in real-time conditions leading to efficient decision-making. In this way smart cities are those cities where investments in human and social capital and modern information and communication technologies (ICT) and its infrastructure fuel sustainable economic growth and a high quality of life, with a wise management of natural resources, through participatory governance (Caragliu and Del Bo, 2011).

These highlight fundamental characteristics of sustainability including energy efficiency, responsible resource management, and citizen engagement (Colldahl, Frey and Kelemen, 2013). For instance, California installed hundreds of thousands of smart metres at homes and commercial establishments that can modulate electricity use with grid supply to prevent loading of infrastructure (Mittal and Sethi, 2018). In light of the growing global environmental challenges, cities need to serve as testing workshops or labs to smartly tackle complex cross-sectional issues like jobs, seamless mobility, safety and security, sustained growth, while responding to the impending climate change too.

There are certain approaches that attempt to articulate smartness into urban planning, governance, and decision-making, for instance the sustainability, metrics, adaptiveness, reporting, technology for inclusiveness (SMART) framework that can be crucial in addressing these multifaceted challenges (Mittal and Sethi, 2018). Smart cities are considered to have six features: smart economy, smart people, smart governance, smart mobility, smart environment, and smart living (Giffinger *et al.*, 2007). The concept is intricately related to intelligent, innovative, or knowledge-based settlements that mobilise ICT to deliver better services, reduce carbon footprint, create sustainable environments, and improve living conditions.

DOI: 10.4324/9781003471417-3

Yet, how cities respond to global climate change in a smart, objective, and systematic manner remains an open question. So, what would make global climate concerns more *integrated* into the smart cities paradigm? A variety of institutions and organisations have a different perspective on integrated approach with respect to climate planning. Some agencies regard integration as a mishmash of various sectors/activities covering both adaptation and mitigation. Within adaptation, integrated essentially considers either vulnerability, impacts, and adaptation or ecological conservation, social development, and disaster risk reduction (Gurung and Bhandari, 2009). Within mitigation, these consider combining reduction of greenhouse gas (GHG) emissions and control of local air pollutants, or linking policies of transport, energy, air pollution, and climate (Creutzig *et al.*, 2019). But this provides an incomplete knowledge of integration, invariably regarding adaptation and mitigation as completely unrelated issues as well as missing out on evidence-based urban climate solutions gathered from case studies and best practices around the world.

Meanwhile, we already know that there is a high degree of fragmentation in urban climate planning raising some crucial knowledge gaps that impede and delay local action, notably non-availability of reliable empirical information on: (a) short, mid-, and long-term climate vulnerability scenarios at sub-national level; (b) how variedly do different societies contribute to change their GHG structures, and (c) what useful climate actions are local governments taking, across the globe (Sethi *et al.*, 2022). These necessitate developing a smart model or tool that integrates such varied but crucial climate concerns of a city into its direct decision-making and long-term planning. In this chapter, we conduct a literature review to have an overview of the state of the affairs on urban climate planning in Asia-Pacific cities (Section 3.2). This is followed by an intensive theoretical understanding on the need of having a smart tool in urban climate action planning (Section 3.3). We then introduce the conceptual framework for Integrated Climate Action Planning (ICLAP) tool, and we establish its applicability in case of Japan (in Section 3.4) discerning urban GHGs, climate vulnerabilities, and identifying relevant urban solutions. The chapter eventually culminates with major scientific findings and policy recommendations (Section 3.5).

3.2 Review of Climate Planning in Asia-Pacific Cities

The review of climate planning in Asia-Pacific cities essentially studies *climate action plans* (CAPs) of 16 major urban centres, covering China (5), Japan (2), India (6), and South-East Asian countries (3). In the case of China, following their 13th Five-Year Plan on Climate Change (2016–2020), cities have a regular system of data inventory of population, economy, urbanisation, energy, and emission indicators, covering adaptation and mitigation aspects of climate change. For instance, Beijing methodically plans to reduce GHGs after 2020 by encouraging building energy efficiency and the use of renewable energy (PGBM – The People's Government of Beijing Municipality, 2016). Another Chinese city, Tianjin, aims to promote carbon trading and modernisation of industrial and energy systems to drastically reduce its emissions (PGTM – The People's Government of Tianjin Municipality, 2017).

The Chinese megacity – Shanghai – has developed both adaptation and mitigation measures into different sectors like industry, energy, construction, and mobility (SMDRC – Shanghai Municipal Development and Reform Commission, 2022). Similarly, Guangzhou and Shenzhen focus on improving climate adaptation in water, cultivation, and maintaining natural systems while also moving towards low-carbon technology (GDRC – Guangzhou Development and Reform Commission, 2017; SDRC – Shenzhen Development and Reform Commission, 2017).

The cities from India have a high degree of variation in data availability for urbanisation, energy, GHG indicators, as well as climate planning at the city level. This is also a matter of definitions used in climate and energy that disrupts the statistics (Sokołowski, 2019). All the Indian cities, even being major economic centres, do not have any definite system of preparing CAP based on scientific inventorisation of their GHG emissions. For instance, Bengaluru lacks its own climate plan, and Chennai's strategy document merely tackles climate resilience. Similarly, though New Delhi has formulated a climate plan with specific adaptation and mitigation measures (including water, energy, urban planning, health, transport, forest, land use/agriculture), the relation of its targets with current and future GHG scenarios is rather uncertain (GNCTD, 2017). It is also observed that most inventories and climate plans are formulated by non-government agencies like TERI in case of Navi Mumbai and PWC for Kolkata (PWC, 2016; TERI, 2013).

Southeast Asia hosts thriving urban centres including Singapore, Bangkok, and Manila covered in this review. Singapore's National Climate Change Secretariat (NCCS) is responsible for the country's climate plan. It indicated that Singapore aims to reduce their emissions intensity by 36% from 2005 levels by reaching GHG emission peak around 2030. Their climate change mitigation and adaptation strategies include improving energy efficiency (in industry, transport, buildings, household, and water and waste management sectors), deploying low-carbon technology, and promoting awareness and action (NCCS, 2016). In addition, previously implemented flood-control projects and integration of green-blue spaces keeps Singapore well-prepared in terms of climate resilience.

Bangkok's climate concerns were finalised through the Bangkok Action Plan on Global Warming 2007–2012. It then elaborated into the Bangkok Master Plan on Climate Change 2013–2023 which clarified the mitigation and adaptation measures. The plan focuses on five main sectors: transport, energy, waste, urban green planning, and adaptation. It states no specific disaster/climate government agency, but rather a collaboration between responsible government offices related to each sector and the Bangkok Metropolitan Administration (BMA). Regular monitoring and evaluation were initially planned, yet there seems to be no public report or announcement. Various mitigation and adaptation strategies were explored in detail and a 14% reduction of GHG levels were forecasted in the BAU scenario for 2020 (BMA, 2012). The BMA will set a new mitigation target for 2030 for its climate change actions in order to fill the gap between the current target under the Master Plan 2020, and consistent with the implementation under the Paris Agreement as well as Thailand's nationally determined contributions to it.

The Philippines' Manila Bay Sustainable Development Master Plan supports the resiliency and sustainability of the Manila Bay area, covering the national capital region of Metro Manila. Its five focal themes embrace ecosystem protection, improvement in water quality, livelihood enhancement, inclusive growth, and disaster risk reduction and climate adaptation. Meanwhile, the Philippines' Climate Change Commission (CCU) extensively covers Metro Manila's Climate Strategy under the National Climate Plan for 2028 (CCU, 2011). With a base level of 2009, it forecasts five yearly emission targets for 2010 up till 2030 along with associated rise in 2020 and 2050 temperatures followed by formulating detailed adaptation and mitigation measures up till 2016, 2022, and 2028. The review of CAPs in Asia-Pacific cities indicates that the process of urban climate planning is highly inconsistent in the Asia-Pacific region, coupled by lack of determined mitigation targets and proper identification of climate strategies by the decision-makers in several instances. The process gets further complicated by inconsistencies and untimeliness in the implementation and monitoring of climate actions. This makes the use of smart, systematic, and integrated tools in urban climate assessment even more important.

In Japan, our review appraised the cities of Tokyo Metropolitan Government (TMG) and the Osaka prefecture Government (OPG). The OPG has a long-term climate plan spanning 10 years, up to 2030, aiming to not just decrease its emission throughput by 40% over 2013 as the base year, but also formulating detailed projects that aim at reduction of climate impacts (OPG, 2021). Similarly, TMG calls for limiting the rise in average global temperature to 1.5°C through a robust Zero-Emission Tokyo strategy by 2050 that includes over 47 items and 82 actions for 17 targets. This strategy is built over TMG's earlier "Carbon-Minus Tokyo" goal to decrease 25% GHG emissions by 2020 (base year: 2000) that thoroughly experimented with the prominent emission trading scheme known as "Cap-and-Trade Programme" (TMG, 2007), encouraging renewables, efficient fuel technologies, and abatement of GHGs in plastic production.

3.3 Smart Tools in Climate Action Planning

Why do cities require smart metrics and tools for decision-making in climate analysis and planning? Climate change in urban areas plays out in a highly complicated manner and requires a systematic approach and tools to assess intersecting linkages. A scientifically informed policy on climate adaptation and mitigation depends on precise and credible evaluation of its environment and societal impacts, including both the costs and benefit perspectives. For example, the central method to measure impact is the alteration in the GHG emission of an activity, quantified in metric tons of gases involved, according to the intensity of involvement. Likewise, for projects and research studies dealing with climate adaptation, climate variability is assessed in terms of precipitation and temperature changes.

During the last decade or so, there has been a marked advancement in preparation and usage of urban climate tools and models. For instance, in Europe, a 300 million-plus strong urban population is represented by 10,774 participants in the Covenant of Mayors. Its cities are sharing climate initiatives through a Good Practices database that documents city profiles, their case studies, videos of accomplished projects, etc. These encompass different sectors: industry, electricity, residential, transport, municipal waste, buildings, lightings, and agriculture together with the categorised for adaptation actions and mitigation targets spanning for 2020–2050 (Covenant of Mayors, 2021). It is an easy-to-use tool for decision-makers and policy planners disseminating a plethora of climate action information about other cities. Globally, Local Governments for Sustainability, commonly known as ICLEI, provides several smart tools for urban climate planning (ICLEI, 2021), like Clean Air and Climate Protection (CACP) software, Adaptation Database and Planning Tool (ADAPT), and HEAT+. Recently, C40 Cities initiated a smart tool that supports cities to better comprehend interlinkages between climate adaptation and mitigation aspects in jointly moderating climate-induced risks (C40 Cities, 2021a). On one hand, mitigation-side tool carries forward prevailing methods of GPC, that is Global Protocol for Community-scale Greenhouse Gas Emission Inventories. The adaptation tool provides cities to self-evaluate their climate risks with Climate Change Risk Assessment Guidance template (C40 Cities, 2021b).

Urban climate assessment tools can broadly be categorised on the basis of their: (a) purpose; (b) method, and (c) sectoral and geographic relevance, as we elaborate further. We illustrate this by drawing examples from multiple studies, though essentially building on theoretical classification propounded in Sethi (2018), a foremost investigation that reports on 44 urban climate metrics and tools, including 27 from mitigation and 17 from adaptation side. In terms of "purpose", there is a definite intention while designing urban climate tools and metrics, either to provide database/information, evaluation, or simulation capabilities. Database- or information-related tools aim to generate greater awareness or fundamental information through urban data or case studies like Synergies in Multiscale Interlinkages of Eco-social Systems (EU, 2014). Meanwhile, evaluation-purposed tools enable experts and the public to use their city data to appreciate the state of affairs, ascertain exact issues, assess the most appropriate policy initiative from numerous given alternatives. A case in point is the Practical Evaluation Tools for Urban Sustainability (PETUS, 2014). Similarly, simulation-centric tools provide users innovative options for modelling in formulating futuristic or alternate situations, conceptualizations, forecasts, scenarios, etc. that assist in greater understanding of impacts for credible policy-making. For instance, a tool called "Synthetic City" enables urban energy systems modelling that covers first-hand planning of a city, its principal activities, socio-economic configuration, energy, and technology choices to meet their needs (SynCity, 2014).

In order to decipher "method", urban climate tools can be categorised as per their primary scientific method: bibliometric, statistical, and spatial (for details refer Sethi and Mittal, 2022). In terms of "sectoral and geographic relevance", urban climate tools and metrics are extremely varied in geographical and sectoral applications. Some tools are dedicated to the evaluation of impacts and benefits in certain mitigation sectors, namely buildings, mobility, or energy. For example, there are several tools assessing supply and demand parameters of urban mobility. Similarly, many tools evaluate different spatial aspects of urban climate including land use, form/density, city-panning, etc. Likewise, different tools are purposed for specific geographical scale, that is region or country. For instance, Urban CLIM enables decision support for urban climate action covering hazards resilience, water, transport, and health sectors in Asia-Pacific cities (APN, 2015). On the other hand, Green Clime Adapt and Urban Adaptation Support tool (EU and EEA, 2014) is customised for policy advice in European context only.

Smart tools need to be specialised in terms of their intended purpose, focused sector, and geographical coverage and find interconnections in grey areas. A case in point is to acknowledge linkages between adaptation and mitigation during target-setting for a city. Like in inconsistencies associated with estimation of urban GHGs, there are numerous gaps that limit the extensive use of smart urban climate tools, spanning conceptual, methodological, empirical, and policy aspects (Sethi, 2018). Conceptual gaps arise out of insufficient scientific knowledge of current theories about climate associations with urban areas, be it through emissions from contributing sectors or climate impacts arising out of variable hazards, vulnerabilities, and risks. Methodological gaps are a consequence of improper assessment of climate vulnerability or city GHGs. Empirical gaps result from insufficient scientific data or reasoning of urban climate impacts or carbon emissions in evidence of prevalent understanding. Policy-governance gaps arise from inadequate knowhow amongst cities around their role and responsibility to effectively plan for climate action. Under these circumstances, a smart tool intended for the purpose of comprehensively guiding urban climate decision-making in a region must be assimilating different functional sectors and scientific methods.

3.4 Materials and Methods

Recent research elucidates how science-based climate planning in urban areas is utterly complex and undefined, mainly augured by three particular approaches (Sethi *et al.*, 2021): (1) Spatial evaluation approach that evaluates climate variability and its hazard propensity to guide mitigation of disasters, needs for climate adaptation, and actions for long-term resilience; (2) statistical modelling that relies on analysis of demographic and economic indicators, prediction of energy consumption and related GHGs to inform prepare policies for low-carbon development; and (3) case studying that indulges in intense qualitative evaluation of best practices from city-specific or peer-city experiences to apprise strategies in comparable conditions. Under such circumstances, reliable evaluation and modelling of multi-dimensional data and information for urban application becomes paramount. Here, we make use of the ICLAP tool, a simulation-based decision-making model that assimilates established methods in climate mitigation, adaptation, and data science (Sethi *et al.* 2022; APN 2021). The model constitutes an integrated approach to guide policy choices keeping in view tailored mitigation scenarios; projected climate variability (temperature and precipitation) for 2030, 2050, and

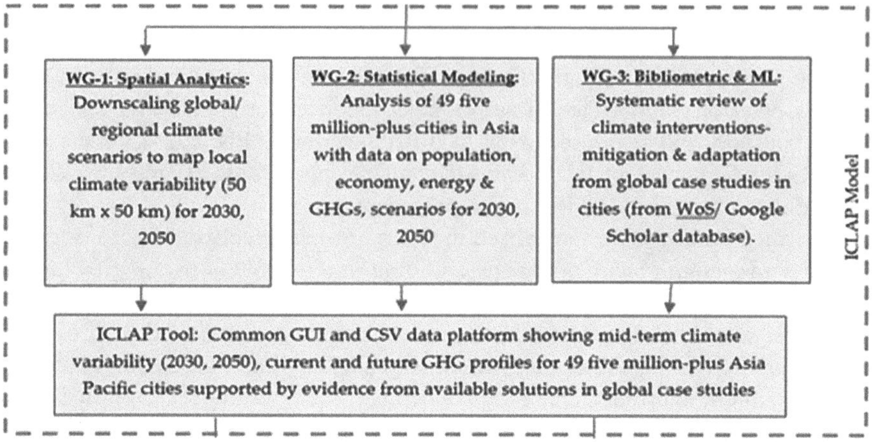

Figure 3.1 The Integrated Climate Action Planning (ICLAP) model.

beyond; and findings from post-factum urban solutions, in a highly organised, transparent, and user-friendly tool interphase.

The ICLAP tool adopts an advanced methodology of synthesising three different knowledge domains/analytics, building on specific and accepted research methods (Figure 3.1) in the respective domains of climate adaptation, mitigation, and data science (Sethi *et al.*, 2022), as summarised later. First, spatial-downscaling climate scenarios and geographic information system (GIS) mapping of variability: In order to forecast climate variabilities at urban region scale, rainfall and temperature deviations from the normal would be downscaled from global/regional MICROC6, with standard regional concentration pathways (RCP) 4.5 and 8.5 scenarios (Saraswat, Kumar and Mishra, 2016) and even up to 2080, eventually being crucial in guiding adaptation alternatives. Using ArcGIS (version 10.3), we perform mapping of rainfall and temperature anomalies of all cities downscaled locally (50 km × 50 km grid) for both 2030 and 2050. The results are reported in the form of shapefile (SHP) along with associated data structure.

Second, statistical trend analysis of urban indicators and GHG forecasts: Climate mitigation assessments would be supported by urban data profiles, their forecasts covering population, economy, energy use in transport, buildings, agriculture/land use, waste, and corresponding GHGs for 2030 and 2050 (Fujimori, Masui and Matsuoka, 2014). The details of relevant datasets used in this step, like World Urbanization Prospects, Global Human Settlement Layer Urban Centres Database (GHSL-UDC), and Emissions Database for Global Atmospheric Research (EDGAR, version 4.3.2) are discussed further. The trends and forecasts (for business as usual, with upper and lower limits) in this step would be instrumental in reviewing current CAPs and identifying new climate mitigation actions.

Third, bibliometric meta-analysis of evidence from case studies: Data extraction and machine learning are employed to systematically review 644 global case studies in local climate action. It employs Google Scholar and Web of Science databases to undertake bibliometric analysis. This is followed by meta-analysing key policy solutions (Lamb *et al.*, 2018, 2019; Sethi *et al.*, 2020), the outputs of which are collated and analysed in standard CSV file format. The results are coded for diverse GHG sectors (energy, industry, transport, land use-land cover change, waste, etc.), their relative efficiency and governance modes for implementation (UN-Habitat, 2011) like regulation, enabling mechanisms, economic instruments, and voluntary measures.

Taking the case study of Indian cities, ICLAP tool is used to test all distinctive features – "database" (in the form of urban solutions), "evaluation" (setting of standard features in downscaled climate variability, GHG scenarios), as well as their customisable "simulation". In this research, we draw from *World Urbanization Prospects: The 2018 Revision*, a standard dataset on urban agglomerations (UNDESA, 2019) for city population. As relevant international conventions, namely Sendai Framework for Disaster Risk Reduction, Sustainable Development Goals, New Urban Agenda and the Paris Agreement (UN, 2015; UNDRR, 2015; UNFCCC, 2015; UN-Habitat, 2016), seek monitoring based on indicators, data of urban area, built-up area, green area, GDP, etc. are drawn from a reliable and uniform source, that is GHSL-UDC (Florczyk *et al.*, 2019). In case of GHG emission data, we know that earlier urban energy/emission studies (Kennedy *et al.*, 2015) performed macroscale correlations (volumetric) and microscale correlations (per capita) for energy use, while Lombardi *et al.* (2017) used similar approach of discerning urban GHGs as "spatial" or "direct", and "economic" or "life-cycle-based" emissions. Thus, to have credible, consistent, and comparable GHG dataset for sampled cities, we draw from European Commission's in-house EDGAR (version 4.3.2), which estimates anthropogenic GHGs from 1970 to 2012 (Crippa *et al.*, 2018). It includes CO_2, NO_x, SO_2 from all production-based activities following standard sector definitions/codes, bottom-up approach of data computation, and hence offering consistency and comparability (IPCC, 1996).

3.5 Results Discussion: The Case of Japanese Cities

This section deliberates on findings from Japan's national climate policy, supported by ICLAP results for two major Japanese cities', Tokyo and Osaka, reported GHG trends and climate variability corresponding to both medium and high scenarios over the 2010s, 2030s, 2050s, and 2080s.

3.5.1 Findings from the National Climate Policy

Japan's national climate policy reflects its commitment to addressing climate change (MoE, 2021). Japan has set targets to achieve net zero GHGs by 2050. This demonstrates a commitment to coping with climate change on a global scale (Sokołowski, 2022). The country has been promoting renewable energy sources like solar and wind power, with a focus on increasing their share in the energy mix. Japan is also focusing on its technological solutions, including artificial intelligence and robotics (Sokołowski, 2021), with its climate policy that encourages innovation and the development of cleaner technologies to reduce emissions, whilst it emphasises improving energy efficiency across sectors, which can lead to reduced emissions and lower energy costs (METI, 2021).

On the other hand, despite efforts to transition to clean energy, Japan continues to rely on fossil fuels, particularly coal and natural gas, which contribute to GHGs (Sokołowski, 2022). The government's support for coal power plants has faced criticism, as coal remains a significant contributor to carbon emissions (Kiko Network, 2018). Its reliance on nuclear power has also raised safety concerns following the Fukushima disaster in 2011, and the debate over nuclear energy remains a contentious issue. Some critics argue that Japan's emission reduction efforts are not sufficient to meet its climate targets, potentially falling short of its 2050 net-zero GHGs goal, with the background that its policy lacks substantial measures and regulations needed to achieve its targets, leaving room for uncertainty and potential loopholes (Renewable Energy Institute, 2020).

The policy at the national government level surely has some influence on local governments in Japan. At the same time, prominent local governments, especially TMG, are regarded as an institution frequently ahead of the national aspiration, in terms of climate mitigation and adaptation actions. In May 2019, Tokyo declared its intent to become a "Zero Emission Tokyo" by 2050 aiding to limit the rise in average global temperature to 1.5°C. It set 17 key targets to be achieved by 2030, including 47 items and 82 actions. Meanwhile, the other megacity of Osaka has a CAP prepared by the OPG with 10 years of planning period, from 2021 to 2030, targeting a GHG emissions reduction by 40% compared with 2013 levels (OPG, 2021). The plan has recently included spatial variabilities of climate change to deliberate on the impact and measures related to adaptation.

3.5.2 Tokyo's GHG and Climate Variability

The GHG emissions value of Tokyo was 128 MtCO_2e in 1975, which escalated to 174 MtCO_2e in 1990 and 225 MtCO_2e in 2015 (ICLAP, 2023). The majority of GHG emissions in 2015 (60%) were from the energy sector (Figure 3.2), followed by industry (21%) and residential sectors (14%). On the other hand, GHG emissions from the transport sector (5%) and agricultural sector (~0%) are negligible. As per the ICLAP model estimates (Figure 3.3), there would be an increase in emissions at 1.4% per annum, leading to 260 MtCO_2e in 2030 and 302 MtCO_2e in 2050.

The results for climate variability in Tokyo indicate a temperature increase of 1.7–4.0°C in the long run (Figure 3.4) depending on the emission scenarios (ICLAP, 2023). The scenario corresponding to the pathway with moderate GHGs (SSP245_MIROC6) exhibits an increase of 1.75°C during 2030s (above the 1980 baseline temperature) to 2°C in 2050s, peaking to 2.75°C by 2080s. The spatial results for moderate-emission scenarios over 2010–2080s are mapped in Figure 3.5. Meanwhile, the scenario corresponding to the pathway with the highest

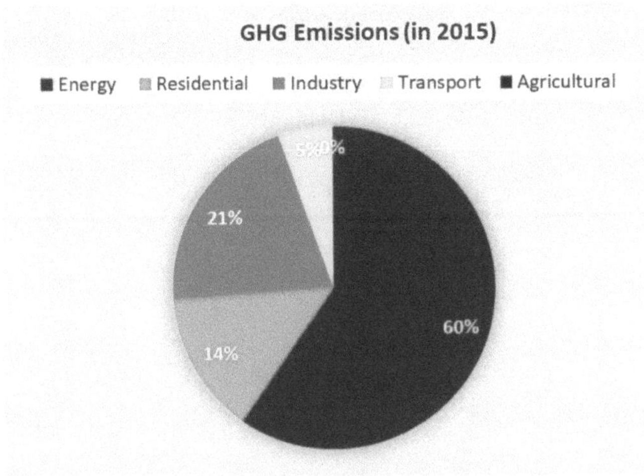

Figure 3.2 GHG contributions from different sectors in Tokyo.

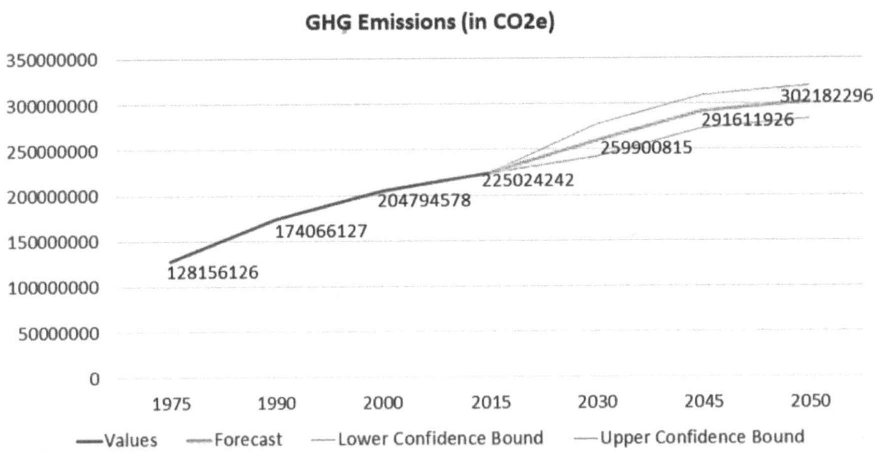

Figure 3.3 ICLAP model estimates for Tokyo's GHG emissions till 2050.

GHGs (SSP585_MIROC6) exhibits an increase of 1.75°C during 2030s (above the 1980 baseline temperature), 2.55°C in 2050s, further consolidating to around 4.0°C above normal during 2080s. The spatial results for high-emission scenarios over 2010–2080s are mapped in Figure 3.6. Meanwhile, the precipitation variation for Tokyo ranges from –100 to 150 mm in the long run (Figure 3.7) depending on the emission scenarios (ICLAP, 2023). The scenario corresponding to the pathway

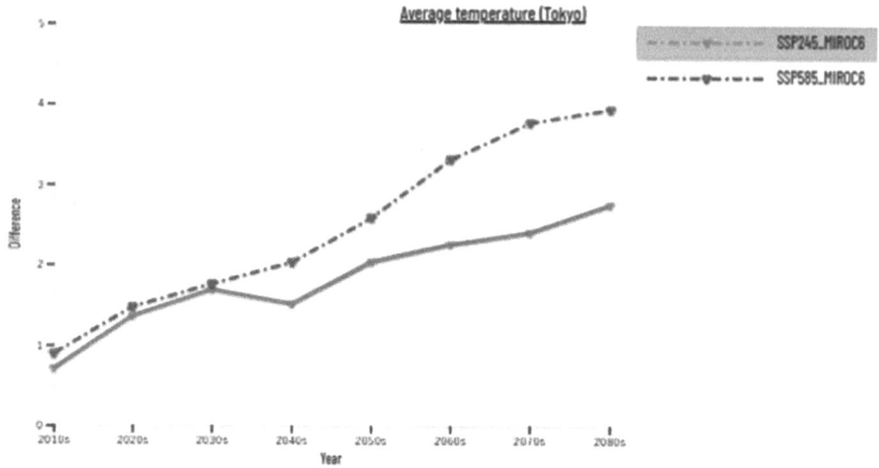

Figure 3.4 Temperature increase in Tokyo under medium- (grey) and high-emission (red) scenario till 2080s.

with moderate GHGs (SSP245_MIROC6) exhibits a rise to 75 mm during 2030s (from the 1980 baseline rainfall), falling back to near normal rainfall in 2050s. This declines further to about –65 mm during 2060s, whereas afterwards it starts re-escalating to 100 mm above 1980 baseline rainfall in the 2080s. The spatial results for moderate-emission scenarios over 2010–2080s are mapped in Figure 3.8. Meanwhile, the scenario corresponding to the pathway with the highest GHGs (SSP585_MIROC6) exhibits a decline of –90 mm in 2030s (from the 1980 baseline rainfall), increasing sharply to 150 mm above normal in the 2040s. This declines significantly to –100 mm below the 1980 baseline rainfall in the 2050s, rising again to 50 mm in 2060s and declining again to –10 mm during the 2080s. The spatial results for high-emission scenarios over 2010–2080s are mapped in Figure 3.9.

3.5.3 Osaka's GHG and Climate Variability

The GHG emission values of Osaka were 62 $MtCO_2e$ in 1975, which escalated to 80.3 $MtCO_2e$ in 1990 and 76.6 $MtCO_2e$ in 2015 (ICLAP, 2023). Most of the emissions in 2015 (41%) were contributed by the energy sector (Figure 3.10), followed by industry (31%). On the other hand, GHG emissions from the residential sector (19%) and transport sector (9%) are relatively minor. As per the ICLAP model estimates (Figure 3.11), there would be an increase in emissions at 0.5% per annum, leading to 85.3 $MtCO_2e$ in 2030 and 91.3 $MtCO_2e$ in 2050.

The results for climate variability in Osaka indicate a temperature increase of 0.7–3.8°C in the long run (Figure 3.12) depending on the emission scenarios (ICLAP, 2023). The scenario corresponding to the pathway with moderate GHGs (SSP245_MIROC6) exhibits an increase of 1.6°C during 2030s (above the 1980

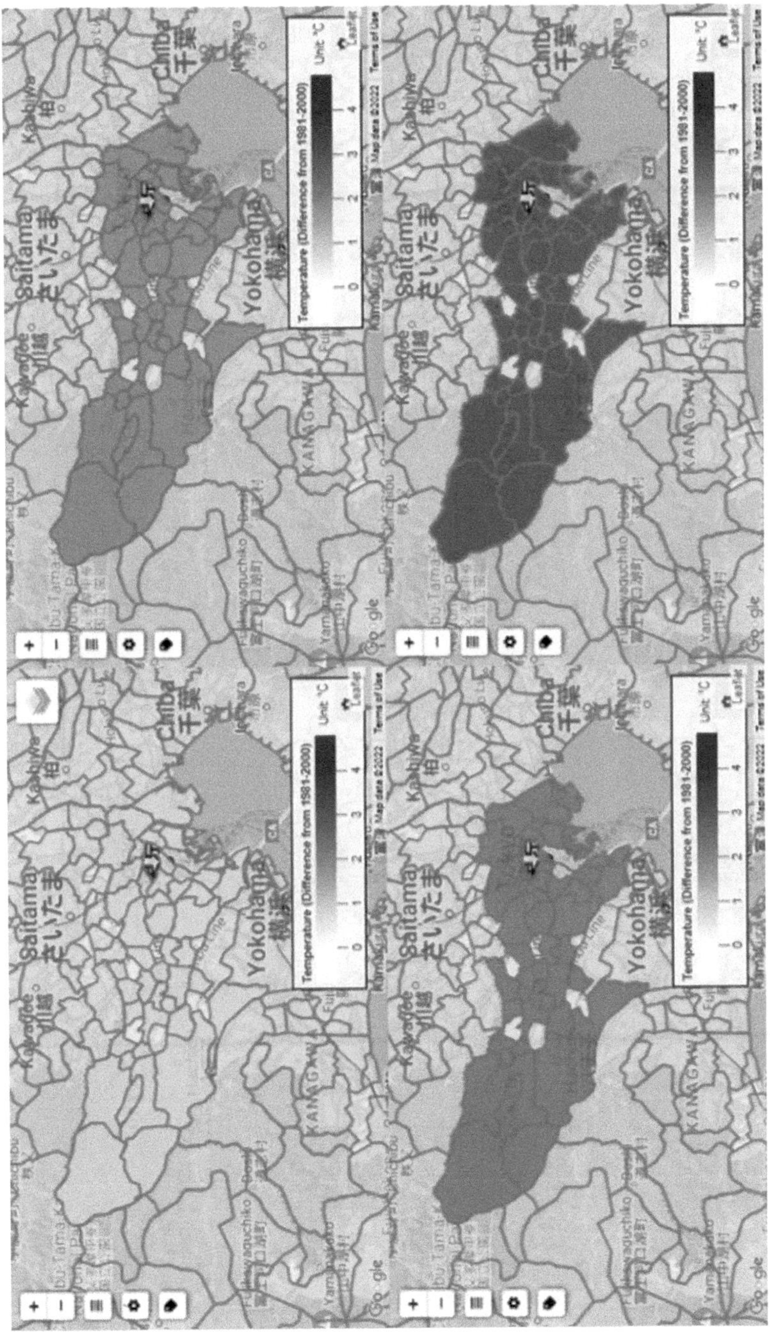

Figure 3.5 Spatial results for Tokyo under medium-emission scenario for 2010s, 2030s, 2050s, 2080s.

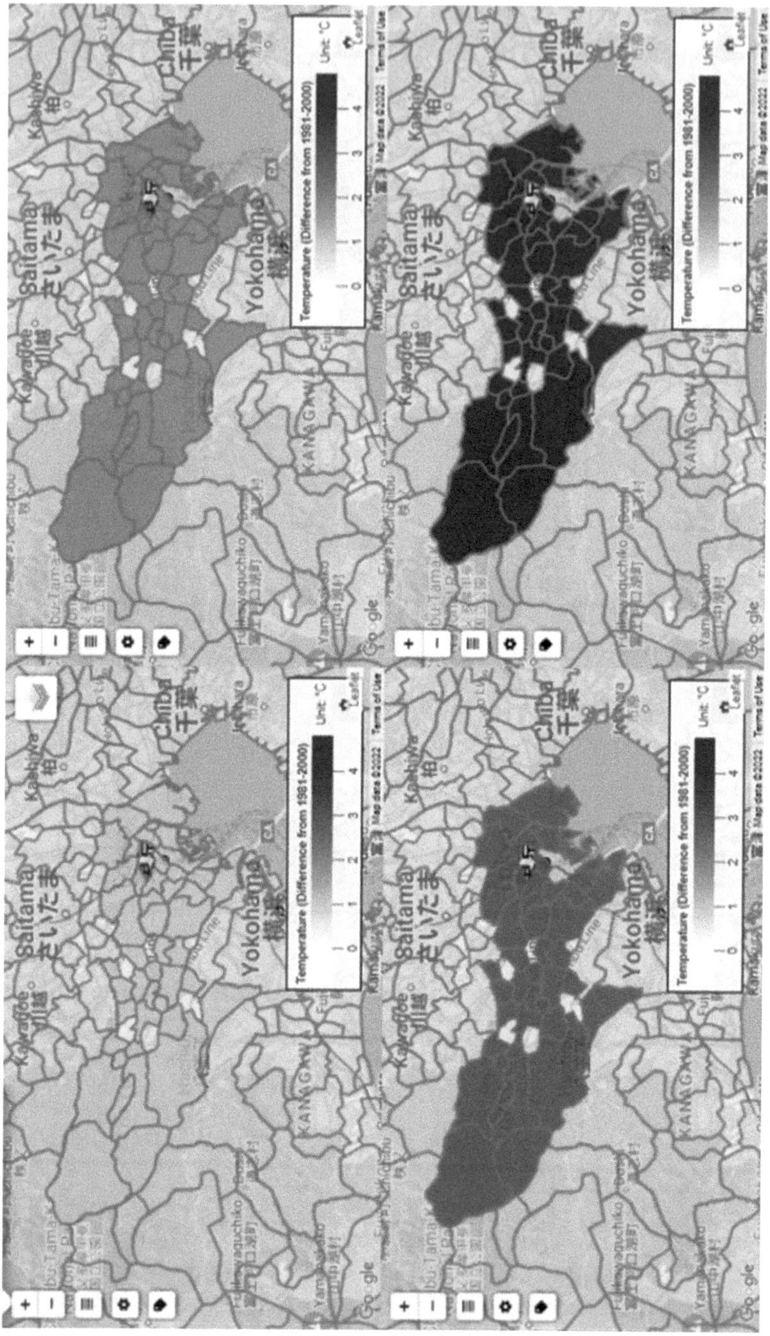

Figure 3.6 Spatial results for Tokyo under high-emission scenario for 2020s, 2030s, 2050s, 2080s.

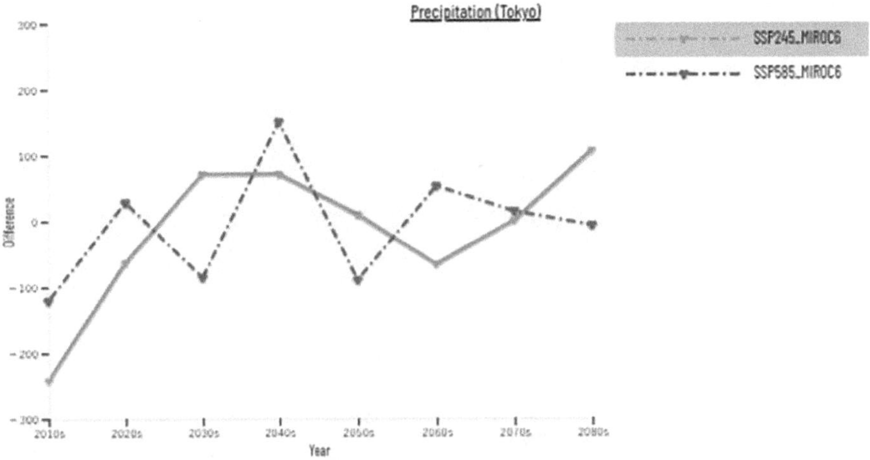

Figure 3.7 Precipitation variation in Tokyo under medium- (grey) and high-emission (red) scenario till 2080s.

baseline temperature) to 2°C in 2050s, peaking to 2.7°C during 2080s. The spatial results for moderate-emission scenarios over 2010–2080s are mapped in Figure 3.13. Meanwhile, the scenario corresponding to the pathway with the highest GHGs (SSP585_MIROC6) exhibits an increase of 1.6°C in 2030s (above the 1980 baseline temperature), 3°C in 2050s, further stabilising to 3.8°C above normal during 2080s. The spatial results for high-emission scenarios over 2010–2080s are mapped in Figure 3.14. Meanwhile, the precipitation variation for Osaka ranges significantly from about –200 to 175 mm in the long run (Figure 3.15) depending on the emission scenarios (ICLAP, 2023). The scenario corresponding to the pathway with moderate GHGs (SSP245_MIROC6) exhibits an increase of 50 mm during 2030s (above the 1980 baseline rainfall), rising to 100 mm in 2040s, diminishing to 130 mm below normal in 2060s, re-escalating to 50 mm above average during 2080s. The spatial results for moderate-emission scenarios over 2010–2080s are mapped in Figure 3.16. Meanwhile, the scenario corresponding to the pathway with the highest GHGs (SSP585_MIROC6) shows there would be negligible precipitation change in 2030s (against the 1980 baseline rainfall), escalating to 170 mm in 2040s, declining again to negligible change from the normal in 2050s, further increasing to 100 mm during 2080s. The spatial results for high-emission scenarios over 2010–2080s are mapped in Figure 3.17.

3.6 Recommendations to Incorporate ICLAP in Smart Cities

Cities bear a strong role and responsibility as agents to mitigate and adapt to climate change. The application of the ICLAP model to cities in Japan demonstrated that both Tokyo and Osaka will face increased rainfall in summer with prolonged

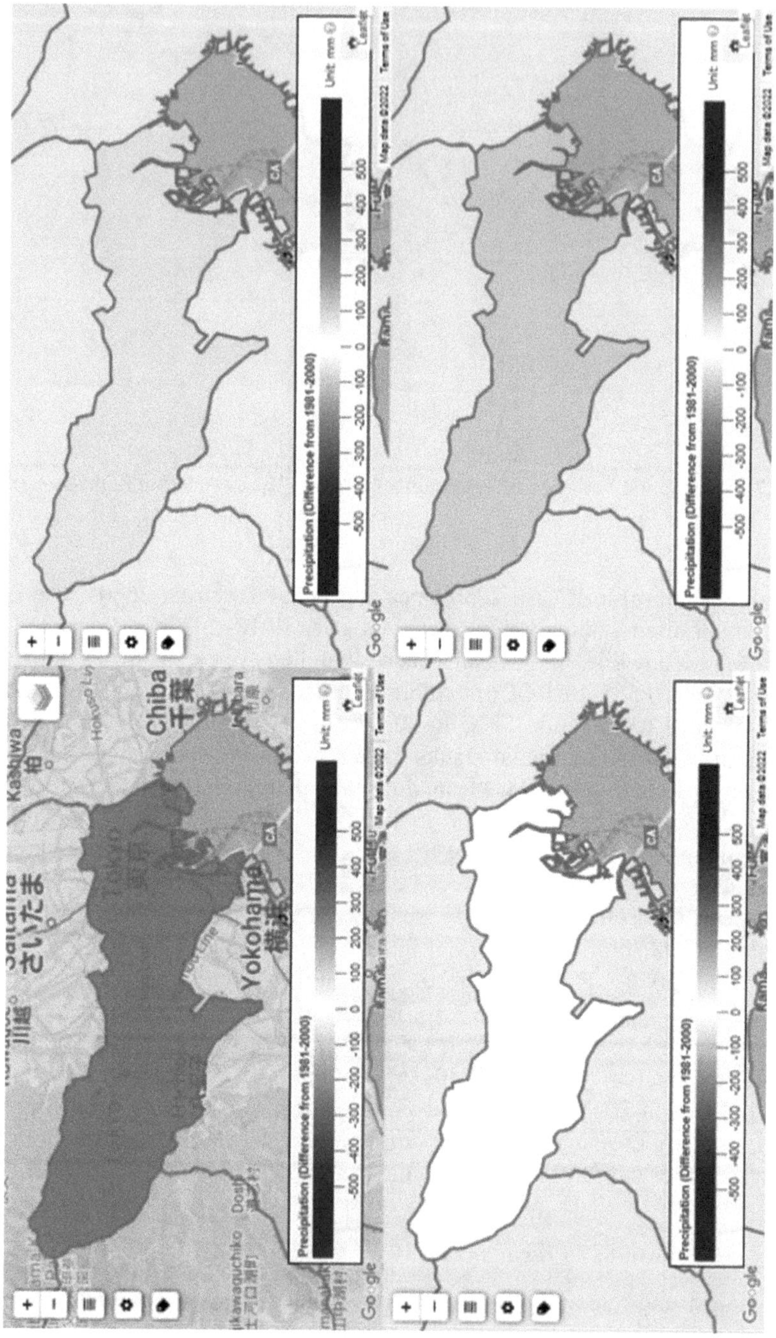

Figure 3.8 Spatial results for Tokyo under medium-emission scenario for 2010s, 2030s, 2050s, 2080s.

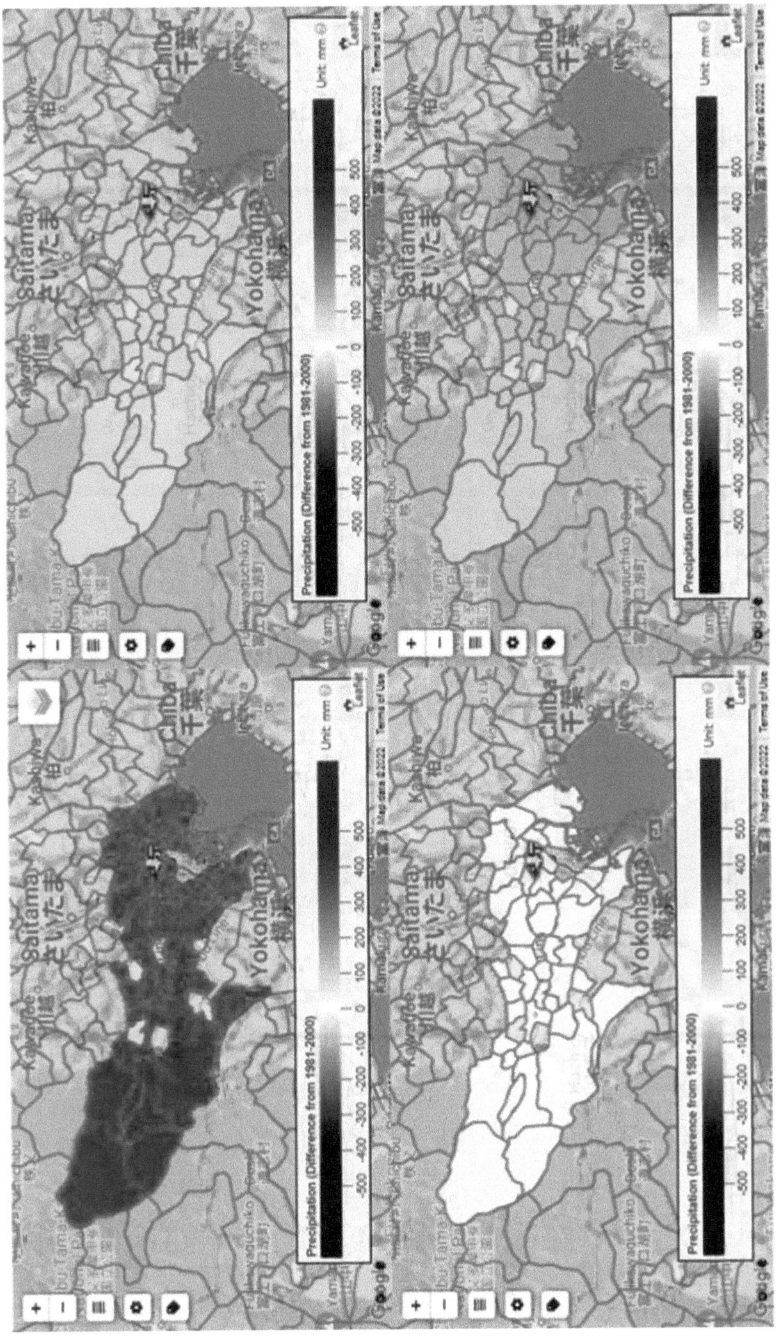

Figure 3.9 Spatial results for Tokyo under high-emission scenario for 2020s, 2030s, 2050s, 2080s.

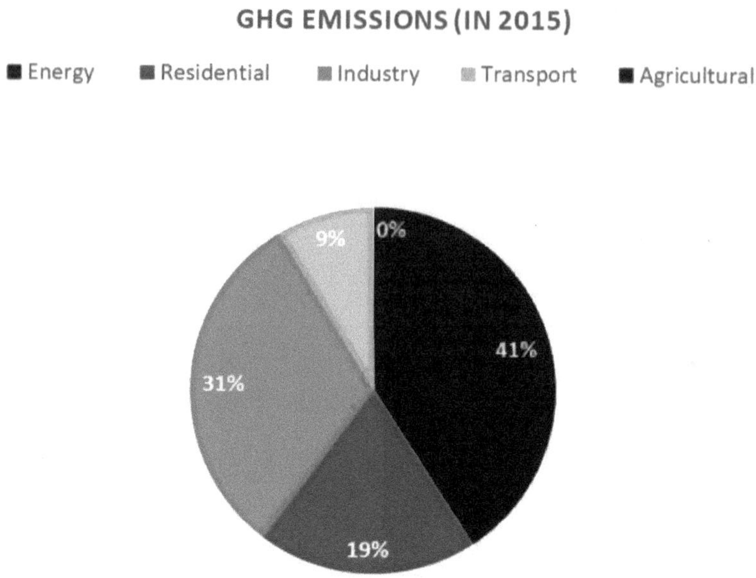

Figure 3.10 GHG contributions from different sectors in Osaka.

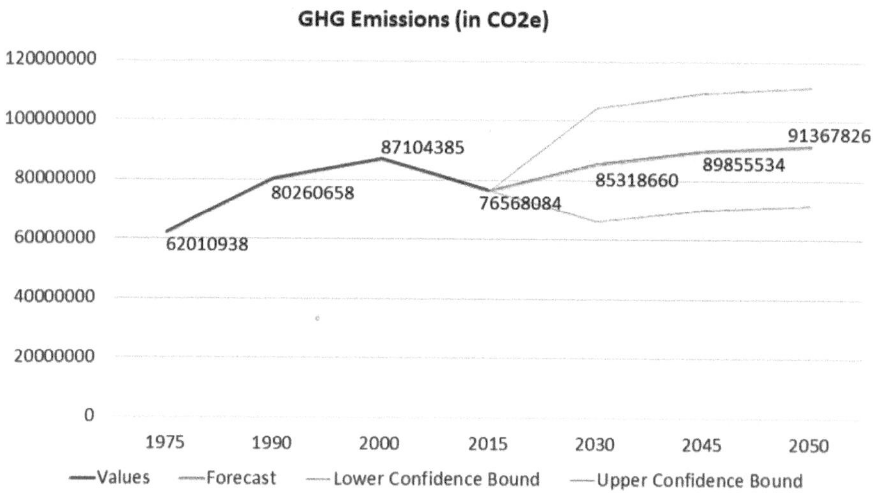

Figure 3.11 ICLAP model estimates for Osaka's GHG emissions till 2050.

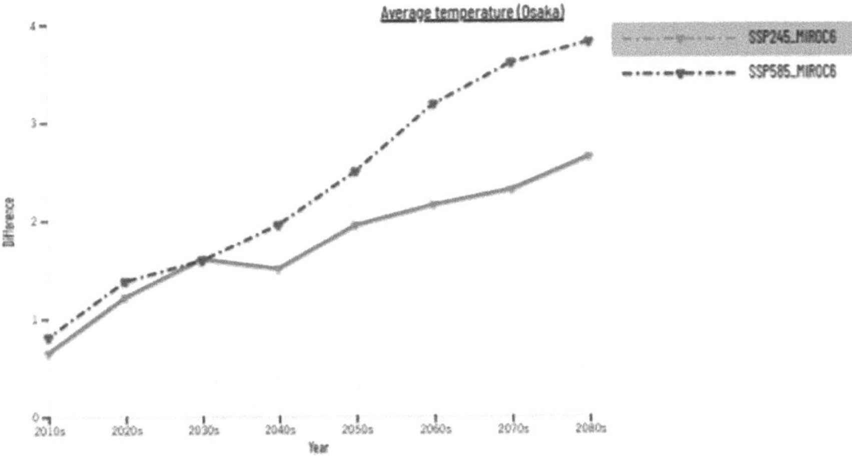

Figure 3.12 Temperature increase in Osaka under medium- (grey) and high-emission (red) scenario till 2080s.

periods of high-humidity days. Stronger adaptive measures are urgently needed to cope with the projected consequences.

The TMG has implemented several climate change adaptation policies to address the growing challenges associated with climate change. It has been proactive in addressing climate change and has recognised the importance of adaptation strategies to protect its residents, infrastructure, and natural environment. Osaka prefecture, like many regions in Japan, has also recognised the importance of climate change adaptation and has been taking steps to address the associated challenges. Some key elements of their climate change adaptation policy of these governments include climate resilience strategy, urban planning and infrastructure, disaster preparedness, flood management, heat stress mitigation, and climate monitoring/research.

With the increased possibility of extreme weather events, however, there is a need to systematically review the current climate mitigation/adaptation policies in relation to the sufficiency of them to meet the projected challenges. This imperative may be better recognised by TMG, which has invested heavily in hydro-infrastructure, for example the floodwater cathedral 22 m underground, as part of the Metropolitan Area Outer Underground Discharge Channel (MAOUDC) being a 6.3 km long system of tunnels and towering cylindrical chambers that protect North Tokyo from flooding (BBC, 2018). Other cities than Tokyo, however, may not be in the similar endowment as Tokyo, in terms of workforce and budget to have the same level of disaster preparedness. Even Tokyo may face adaptive difficulties when the rainfall and humidity increase as forecasted.

The Tokyo and Osaka governments have the potential to increase the renewable energy mix while maintaining supply reliability. Strategic planning, grid

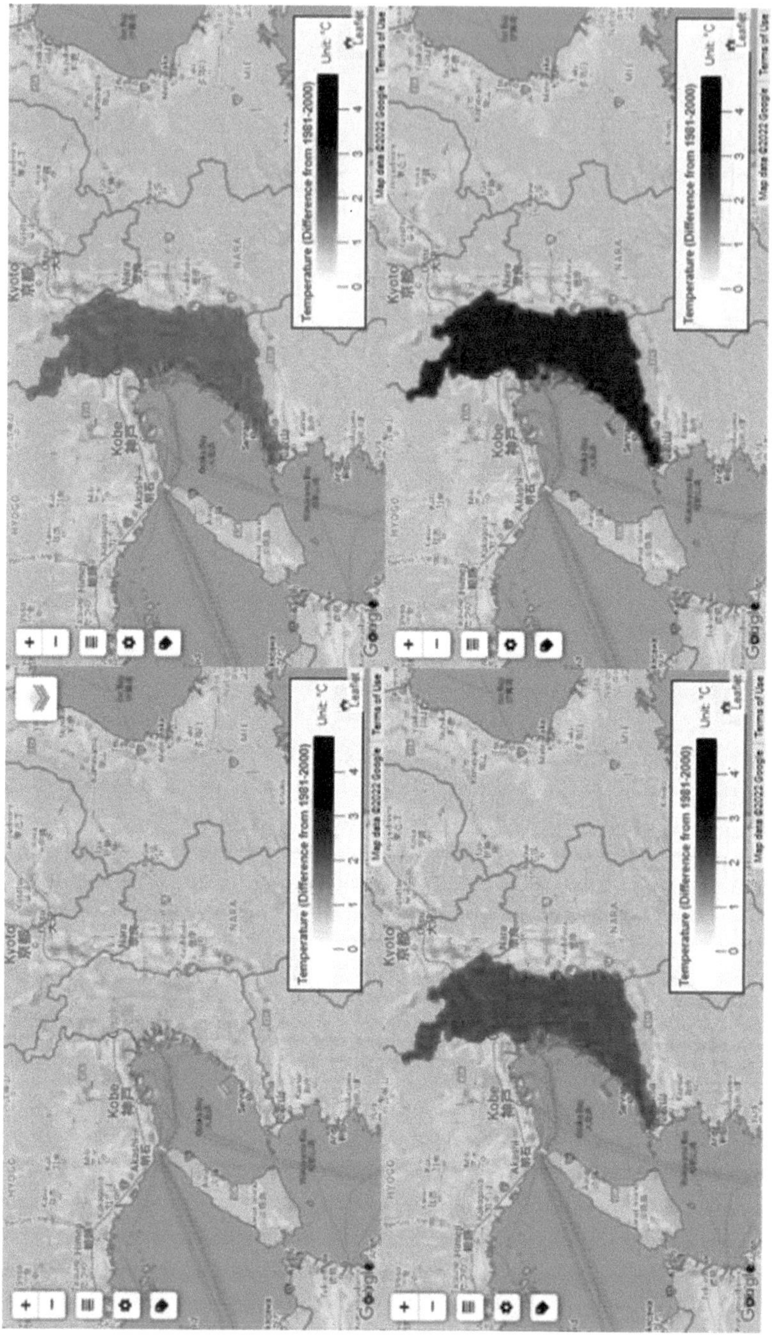

Figure 3.13 Spatial results for medium-emission scenario for 2010s, 2030s, 2050s, 2080s.

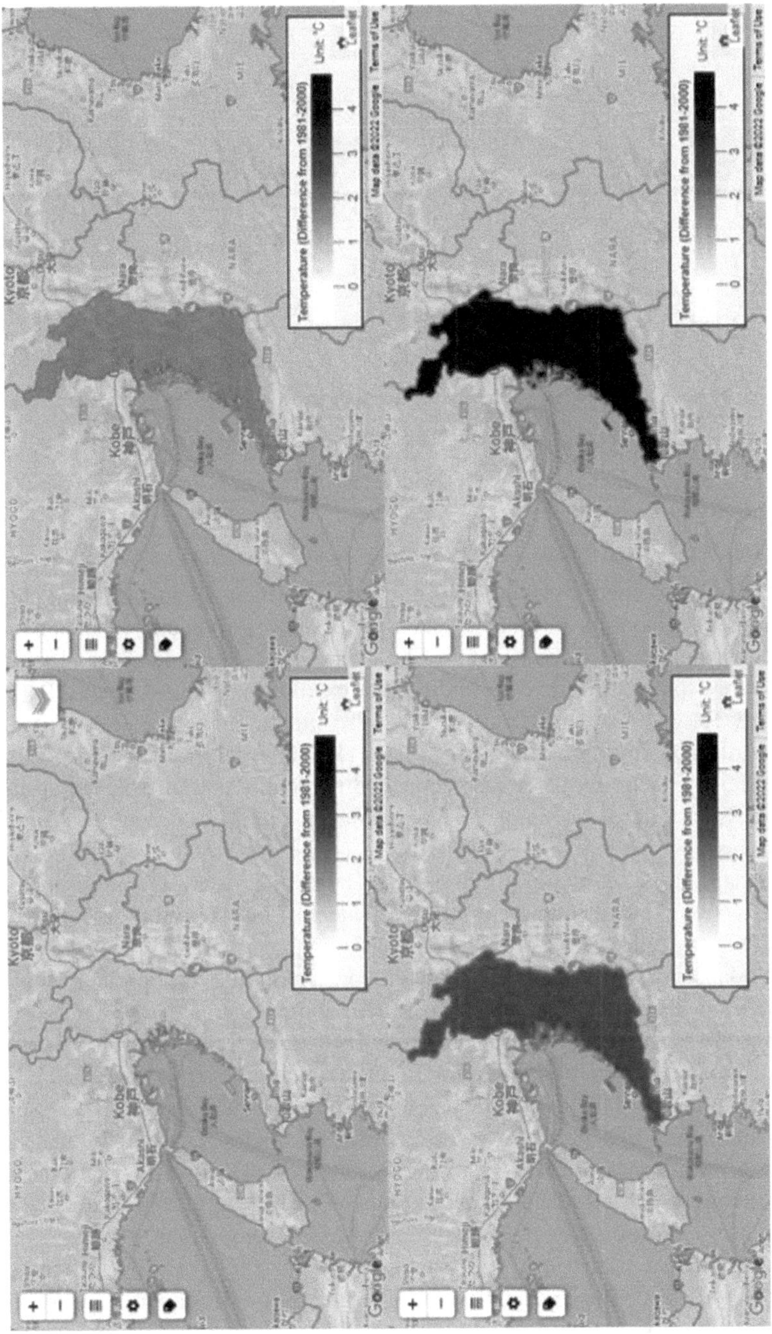

Figure 3.14 Spatial results for high-emission scenario for 2020s, 2030s, 2050s, 2080s.

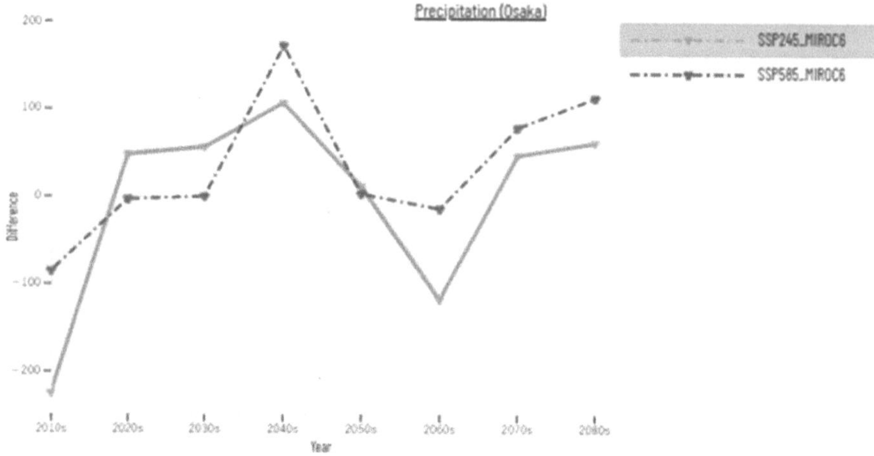

Figure 3.15 Precipitation variation in Osaka under medium- (grey) and high-emission (red) scenario till 2080s.

enhancements, and energy storage investments can mitigate the intermittent nature of renewables. Tokyo is moving ahead to introduce provisions to make solar installations mandatory for new homes. The rules apply to homes with total rooftop areas of more than 20 m^2, and to buildings with rooftops smaller than 2,000 m^2 from 2025. Tokyo's Cap and Trade scheme also functions well to ensure gradual reduction of CO_2 emissions in their domain (Wakabayashi and Kimura, 2018). Smart policies encouraging private sector and residents' participation in sustainable projects and incentivising clean energy adoption will be continuously crucial to both cities.

At the same time, contemporary research shows that urban climate issues are particularly fraught with complexity due to intersection of large number of distinct variables in "climate geoscience" (precipitation and temperature anomalies at different locations, RCPs, timeline), "adaptation alternatives" (approach, priority, intervention level), and "urban governance" (functional mandate, institutional capacity, as well as plans and policies) and require systematic negotiation process to situate, ground, and operationalise climate action in cities (Sethi *et al.*, 2021). This necessitates transition to multi-dimensional and multi-agent interactive systems. The transition requires interactions of multiple technologies, stakeholders, and governance frameworks (Sokołowski and Visvizi, 2023), also to be more context dependent, breaking through the burdens of path dependency (Sethi, 2022). It is meant to overcome the short-termism currently dominating urban planning, and instead embrace long-term integrated global perspectives. To implement it through climate governance, the mixture of adaptive, precautionary, and reactive policies becomes imperative. Precautionary policies are necessary to limit harmful surprises but due to the current trends of change it is inevitable to prepare for

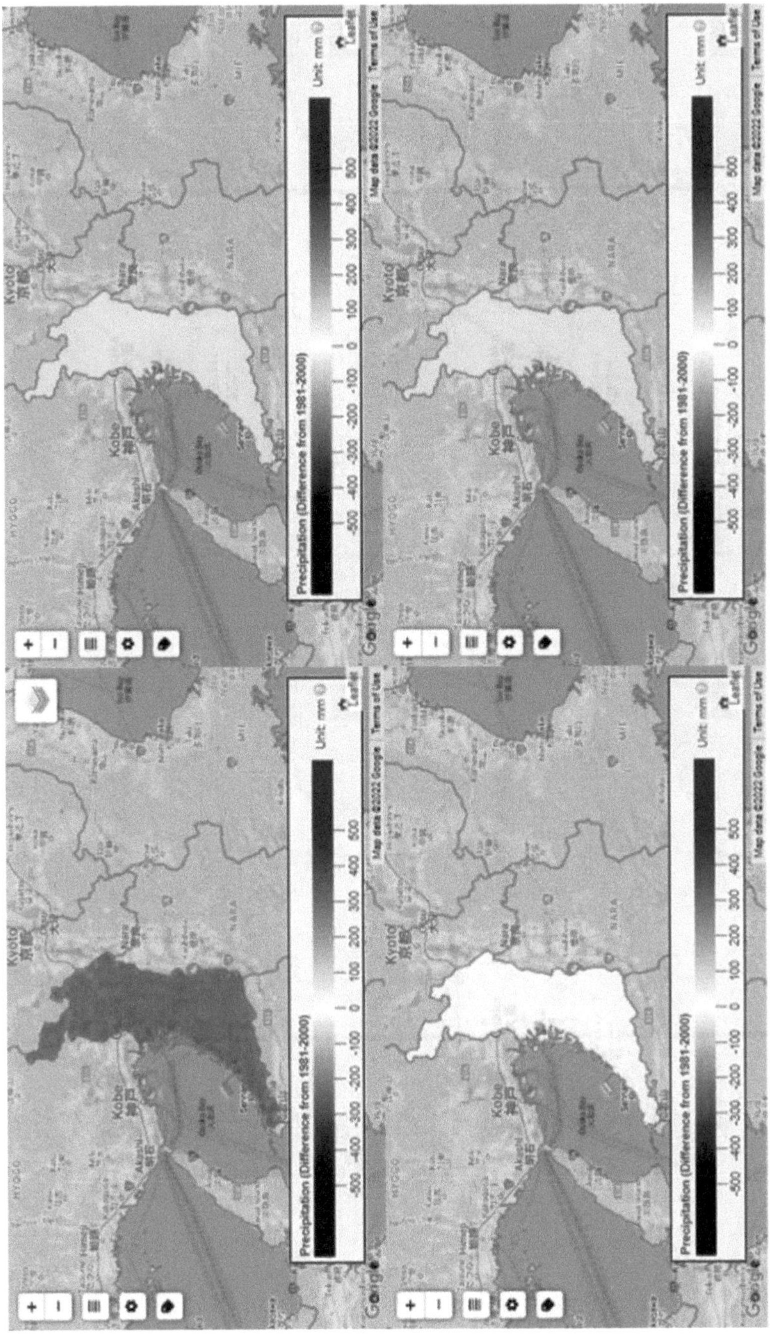

Figure 3.16 Spatial results for medium-emission scenario for 2010s, 2030s, 2050s, 2080s.

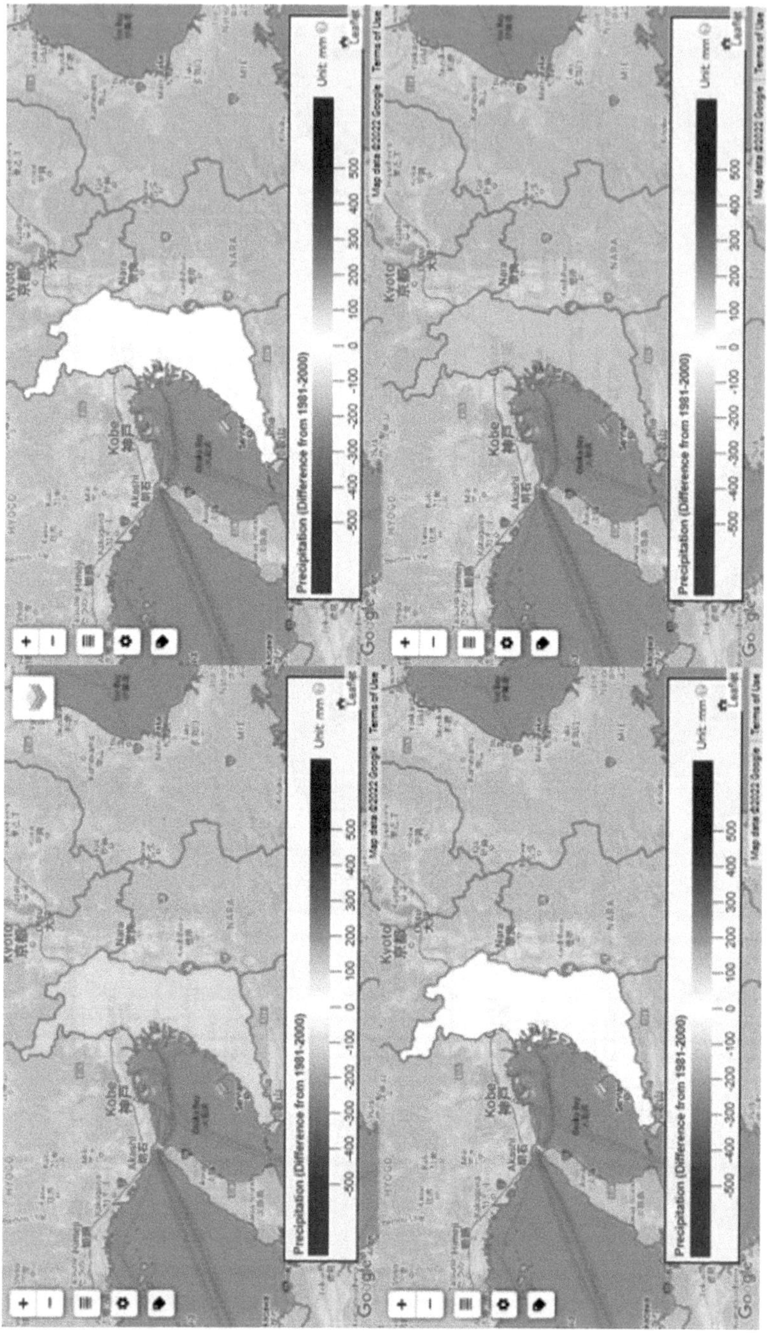

Figure 3.17 Spatial results for high-emission scenario for 2020s, 2030s, 2050s, 2080s.

system changes. Therefore, policies based on scientific tools like ICLAP become necessary to increase the adaptive capacities in the environment and that of the urban societies.

Similarly, developing smart metrics and evaluation tools for climate action is an intense process of scientific collaboration, data collection, knowledge sharing, identification of tools and methods, to implementation and policy review. It requires comprehensive knowledge of multiple disciplines, existing methodologies, and tools to effectively integrate these ensuring greater field application. Integrated models provide an opportunity for partnership between interdisciplinary experts and city managers to facilitate a platform for exchanging knowledge that can ultimately build capacities for technical and local governance processes. In the long term, ICLAP-like metrics and tools act as an instrument of science-based policy application in local urban and sub-national development, cooperating with the private sector and encouraging public participation. Given the practical relevance, ICLAP should be incorporated within national urban sustainability initiatives and their associated funding mechanism with state governments, guiding stakeholders in sensible target setting, prioritising projects, and tracking their progress. Since the phenomenon has global causations and consequences, using such urban climate tools will also promote comparisons with other cities, interactions with intercontinental city forums, scientist community, non-governmental organisations, businesses, and general public. In addition, the resulting multi-sectoral strategies would foster inter-departmental coordination, enhance public awareness on climate science, and promote private and civic action in urban systems. In this regard, national governments as well as regional and international scientific bodies should mobilise greater support to integrated and collaborative scientific tools showing significant practical applicability on the ground to combat global climate challenge.

Acknowledgments

The authors would like to acknowledge Li-jing Liu, Eva Ayaragarnchanakul, Akhilesh Surjan, and Ram Avtar for their preliminary data inputs about climate action plans in Asia-Pacific cities.

Funding

The first author acknowledges the grants from the Asia-Pacific Network for Global Change Research under their Collaborative Regional Research Programme (CRRP), Project No. CRRP2020-04MY-Sethi, titled "Integrated climate action planning (ICLAP) 2050 tool in Asia-Pacific cities".

Consent for Publication

This theoretical basis of the chapter (sections 3.1–3.4) essentially draws from the article "Developing a smart tool for integrated climate action planning (ICLAP 2050) in Asia-Pacific Cities", authored by Mahendra Sethi and Shilpi Mittal and published in *Computational Urban Science* (2022). To preserve the thematic focus of this chapter, its Section 3.3 on "Smart Tools in Climate Action Planning" offers a slightly abridged version of the original article.

References

APN. (2015). "*UrbanCLIM*. Asia Pacific Network". Available at: www.apn-gcr.org/resources/items/show/1749

APN (2021) "Integrated Climate Action Planning (ICLAP) 2050 Tool in Asia-Pacific Cities". Asia-Pacific Network for Global Change Research.

BBC (2018) "The Underground Cathedral Protecting Tokyo from Floods". Available at: www.bbc.com/future/article/20181129-the-underground-cathedral-protecting-tokyo-from-floods (Accessed: 29 March 2024).

BMA (2012) "Bangkok Master Plan on Climate Change 2013–2023". Available at: www.start.chula.ac.th/start/images/Champ/Day1/Bangkok-Master-Plan-on-Climate-Change_31July2017.pdf (Accessed: 29 March 2024).

C40-Cities (2021a) "New Tool Will Help Cities Understand Interactions between Mitigation and Adaptation Actions". Available at: www.c40.org/news/new-tool-will-help-cities-understand-interactions-between-mitigation-and-adaptation-actions/.

C40-Cities (2021b) "Assessing Risks in Cities". Available at: https://resourcecentre.c40.org/resources/assessing-risks-in-cities.

Caragliu, A. and Del Bo, C.F. (2019) "Smart Innovative Cities: The Impact of Smart City Policies on Urban Innovation", *Technological Forecasting and Social Change*, 142, pp. 373–383.

CCU (Climate Change Commission) (2011) "National Climate Change Action Plan 2011–2028". Available at: http://climate.emb.gov.ph/wp-content/uploads/2016/06/NCCAP-1.pdf (Accessed: 29 March 2024).

CIDOB (2014) "Beyond Smart Cities: It's Time for Urban Sustainable Development", *Notes Internacionals CIDOB*, 92, pp. 1–5.

Colldahl, C., Frey, S. and Kelemen, J.E. (2013) *Smart Cities: Strategic Sustainable Development for an Urban World*. Karlskrona: Blekinge Institute of Technology.

Covenant of Mayors (2021) "Plans & Actions". Available at: www.covenantofmayors.eu/plans-and-actions/good-practices.html (Accessed: 29 March 2024).

Creutzig, F. *et al.* (2019) "Upscaling Urban Data Science for Global Climate Solutions", *Global Sustainability*, 2, p. e2.

Crippa, M. *et al.* (2018) "Gridded Emissions of Air Pollutants for the Period 1970–2012 Within EDGAR v4. 3.2", *Earth System Science Data*, 10(4), pp. 1987–2013.

EU (2014). "Sustainable Urban Transport Policies and Initiatives (SMILE)". Available at: http:// ec.europa.eu/environment/urban/urban_transport.htm.

EU and EEA (2014). "Urban Adaptation Support Tool". Available at: http://climate-adapt.eea. europa.eu/tools/urban-ast/step-0-0.

Florczyk, A.J. *et al.* (2019) "Description of the GHS Urban Centre Database 2015", *Public Release*, 1, pp. 1–75.

Fujimori, S., Masui, T. and Matsuoka, Y. (2014) "Development of a Global Computable General Equilibrium Model Coupled with Detailed Energy End-Use Technology", *Applied Energy*, 128, pp. 296–306.

GDRC (2017) "Guangdong's 13th Five-Year Plan on Climate Change". Guangzhou Development and Reform Commission.

Giffinger, R. *et al.* (2007) "City-Ranking of European Medium-Sized Cities", *Centre of Regional Science at the Vienna University of Technology*, 9(1), pp. 1–12.

GNCTD (2017) "Delhi State Action Plan on Climate Change". Available at: http://moef.gov.in/wp-content/uploads/2017/08/Delhi-State-Action-Plan-on-Cimate-Change.pdf (Accessed: 29 March 2024).

Gurung, G.B. and Bhandari, D. (2009) "Integrated Approach to Climate Change Adaptation", *Journal of Forest and Livelihood*, 8(1), pp. 91–99.

ICLAP (2023) "Integrated Climate Action Planning Tool". *Indian Society for Applied Research and Development*. Available at: www.iclap2050.in/japan (Accessed: 29 March 2024).

ICLEI (2021) "List of Tools". Available at: http://old.iclei.org/index.php?id=19 (Accessed: 29 March 2024).

IPCC (1996) "IPCC Guidelines for National Greenhouse Gas Inventories".

Kennedy, C.A. *et al.* (2015) "Energy and Material Flows of Megacities", *Proceedings of the National Academy of Sciences*, 112(19), pp. 5985–5990.

Kiko Network (2018) "Japan Coal Phase-Out: The Path to Phase-Out by 2030". Available at: www.kikonet.org/kiko/wp-content/uploads/2018/11/Report_Japan-Coal-Phase-Out_EG.pdf (Accessed: 29 March 2024).

Lamb, W.F. *et al.* (2018) "The Literature Landscape on 1.5 C Climate Change and Cities", *Current Opinion in Environmental Sustainability*, 30, pp. 26–34.

Lamb, W.F. *et al.* (2019) "Learning about Urban Climate Solutions from Case Studies", *Nature Climate Change*, 9(4), pp. 279–287.

Lombardi, M. *et al.* (2017) "Assessing the Urban Carbon Footprint: An Overview", *Environmental Impact Assessment Review*, 66, pp. 43–52.

METI (2021) "Overview of Japan's Green Growth Strategy: Through Achieving Carbon Neutrality in 2050".

Mittal, S. and Sethi, M. (2018) "Smart and Livable Cities: Opportunities to Enhance Quality of Life and Realize Multiple Co-Benefits", in M. Sethi and J.A. Puppim de Oliveira (eds) *Mainstreaming Climate Co-Benefits in Indian Cities: Post-Habitat III Innovations and Reforms*. Singapore: Springer, pp. 245–263.

MoE (2021) "Achieving Net Zero GHG Emissions by 2050 in Japan: Policies and Measures Update". Available at: www.mofa.go.jp/files/100153687.pdf (Accessed: 29 March 2024).

NCCS (2016) "Take Action Today, for a Carbon-Efficient Singapore".

OPG (2021) "Osaka Prefecture Global Warming Countermeasures Implementation Plan (Area Measures)".

PETUS (2014) "Practical evaluation tools for urban sustainability". http://www.petus.eu.com.

PGBM (2016) "Plan for Energy Saving and Consumption Reduction and Climate Change Addressing in Beijing during the 13th Five-Year Plan Period".

PGTM (2017) "Implementation Plan of Greenhouse Gas Emission Control in Tianjin During The 13th Five-Year Plan Period".

PWC (2016) "Roadmap for Low Carbon and Climate Resilient Kolkata".

Renewable Energy Institute (2020) "5 Fallacies of Japan's Coal-fired Power Export Policy". Available at: www.renewable-ei.org/pdfdownload/activities/202003_coalexpinfo_EN.pdf (Accessed: 29 March 2024).

Saraswat, C., Kumar, P. and Mishra, B.K. (2016) "Assessment of Stormwater Runoff Management Practices and Governance under Climate Change and Urbanization: An Analysis of Bangkok, Hanoi and Tokyo", *Environmental Science & Policy*, 64, pp. 101–117.

SDRC (2017) "Shenzhen's 13th Five-Year Plan on Climate Change".

Sethi, M. (2018) "Co-Benefits Assessment Tools and Research Gaps", in M. Sethi and J.A. Puppim de Oliveira (eds) *Mainstreaming Climate Co-Benefits in Indian Cities: Post-Habitat III Innovations and Reforms*. Singapore: Springer, pp. 47–73.

Sethi, M. (2022) *Sustainable Societies: Transition from Theories to Practice*. Berlin: Technical University of Berlin.

Sethi, M. and Mittal, S. (2022) "Developing a Smart Tool for Integrated Climate Action Planning (ICLAP 2050) in Asia-Pacific Cities", *Computational Urban Science*, 2(1), p. 45.

Sethi, M. *et al.* (2020) "Climate Change Mitigation in Cities: A Systematic Scoping of Case Studies", *Environmental Research Letters*, 15(9), p. 093008.

Sethi, M. *et al.* (2021) "How to Tackle Complexity in Urban Climate Resilience? Negotiating Climate Science, Adaptation and Multi-Level Governance in India", *PLOS One*, 16(7), p. e0253904.

Sethi, M. *et al.* (2022) "Integrated Climate Action Planning (ICLAP) in Asia-Pacific Cities: Analytical Modelling for Collaborative Decision Making", *Atmosphere*, 13(2), p. 247.

SMDRC (2022) "Shanghai's Key Work Arrangements for Energy Conservation, Emission Reduction and Climate Change in 2020". Shanghai Municipal Development and Reform Commission.

Sokołowski, M.M. (2019) "When Black Meets Green: A Review of the Four Pillars of India's Energy Policy", *Energy Policy*, 130, pp. 60–68.

Sokołowski, M.M. (2021) "Artificial Intelligence and Climate-Energy Policies of the EU and Japan", in D. Bielicki (ed.), *Regulating Artificial Intelligence in Industry*. London: Routledge, pp. 138–155.

Sokołowski, M.M. (2022) *Energy Transition of the Electricity Sectors in the European Union and Japan: Regulatory Models and Legislative Solutions*. Palgrave Macmillan: Cham.

Sokołowski, M.M. and Visvizi, A. (2023) "Exploring the Smart Cities – Energy Communities Nexus", in M.M. Sokołowski and A. Visvizi (eds) *Routledge Handbook of Energy Communities and Smart Cities*. London: Routledge, pp. 1–10.

SynCity (2014) "Synthetic City for Modeling Urban Energy Systems". Available at: http://secon dlife.com/destination/syn-city (Accessed: 29 March 2024).

TERI (2013) "Carbon Inventory of Navi Mumbai".

TMG (2007) "Tokyo Climate Change Strategy: Progress Report and Future Vision".

UN (2015) "The Sustainable Development Goals".

UNDESA (2019) "World Population Prospects 2019: Highlights".

UNDRR (2015) "The Sendai Framework for Disaster Risk Reduction 2015–2030".

UNFCCC (2015) "The Paris agreement".

UN-Habitat (2011) "Cities and Climate Change: Global Report on Human Settlements".

UN-Habitat (2016) "The New Urban Agenda".

Wakabayashi, M. and Kimura, O. (2018) "The Impact of the Tokyo Metropolitan Emissions Trading Scheme on Reducing Greenhouse Gas Emissions: Findings from a Facility-Based Study", *Climate Policy*, 18(8), pp. 1028–1043.

4 5Grids Concept as a Basis of Zero-Carbon, Decentralised, and Digitalised Society

Hideo Ishii and Yasuhiro Hayashi

4.1 Introduction

The Intergovernmental Panel on Climate Change (IPCC) has continued to communicate scientific knowledge that global warming is actually progressing, with the negative impacts it brings. Coupled with the unusual weather conditions and frequent occurrence of natural disasters in various parts of the world, the international community has begun to move towards carbon neutrality as the immediate major goal for all humanity. The acclaimed basis for achieving carbon neutrality consists of renewable energy sources, reduction of emissions, and improved energy use.

With respect to energy efficiency, Japan has a vast experience in developing highly energy-efficient equipment, such as electric vehicles and water heaters that use heat pumps or fuel cells. In addition, research and development of an energy management system (EMS) that optimally operates these devices in an integrated manner has been carried out vigorously in Japan (Ishii *et al.*, 2016). Due to the benefits of connecting the EMS with various devices like data exchange, Japan has been promoting the standardisation of communication interfaces to optimise energy management and its extensions like demand response (DR) and virtual power plants (VPP) (Ishii and Hirohashi, 2019).

Additionally, the introduction of smart electricity metres is progressing in countries around the world including Japan. This makes it possible to collect digital data on electricity consumption in units of 15–30 minutes. The data can be used for various purposes such as analysis of consumption patterns, implementation of energy-saving measures, introduction of time-of-use charges, dynamic pricing, and DR. The digitisation of such data is progressing rapidly in all areas, and smart metres are also being introduced for gas and water consumption. Furthermore, in recent years, the use of global positioning system (GPS) has become more widespread, collecting real-time data with the movements of mobile phones and cars. Use of this data makes it possible to grasp information on the flow of people and the state of road congestion at a glance. It is becoming possible to understand in detail what kind of products are connected to the Internet, and when and where they are being used. It seems inevitable that the use of these kinds of big data will rapidly advance as computers, internet of things (IoT), and artificial intelligence (AI) become more sophisticated.

DOI: 10.4324/9781003471417-4

On the other hand, due to the lack of established mechanisms and rules for sharing data, it is still limited to use for its own primary purpose. It can be said that we are at the stage where various buds are beginning to appear, leading to a world in which data from different fields are combined, analysed, and utilised. With this in mind, this chapter is focused on five components of public infrastructures: electricity, gas, water, transportation, and information, and initiated research aimed at collecting various elements of infrastructural big data to integrate into a data collaboration platform that enables various applications. This represents our concept of 5Grids with a goal to provide a basis for building future smart cities.

In the following sections, we describe the origins of Japan's energy transition (Section 4.2), and then discuss the current state and prospect for smart metres of electricity in Japan (Section 4.3). Section 4.4 presents the research and development of distributed cooperative EMS tool that we have been working on as a tool for smart cities. Section 4.5 explains the concept of 5Grids. Section 4.6 introduces efforts to build a smart city based on sector coupling of electricity and transportation in practice, which is a preliminary application of 5Grids concept in real life. Finally, the challenges and prospects concerning 5Grids concept are highlighted in the summary.

4.2 Energy Transition in Japan and Innovations on Demand Side

Japan is poor in energy resources and relies heavily on imports from overseas. Building good relations with other countries and ensuring energy security remains a top policy priority. Japan has been diversifying its energy resources from the period of high economic growth to the present. During the 1970s, the oil crisis led to a shift to natural gas and the introduction of nuclear power, and from the late 1990s to the 2000s, when debate about preventing global warming intensified, efforts were made to reduce carbon in power sources, such as by technological developments in the thermal power generation, including combined heat and power (Sokołowski, 2022a, pp. 189–197). As a result, Japan has achieved an overall thermal efficiency of 60% for natural-gas-fired power generation (Sano, 2011).

Another major characteristic of Japan is its efforts to improve energy efficiency and save energy on the energy consumption side. Japan has a long-standing tradition of thriftiness and a deep-rooted cultural orientation towards saving energy (Humans, 2023). Indeed, various efforts have been made to create advanced production technologies for Japanese industries, and with the help of research and development the energy consumption efficiency on the demand side has been significantly improved. As a result, Japan has built high-performance equipment and systems like batteries, heat pumps, fuel cells, hydrogen production and utilisation facilities, as well as implemented energy management at various levels, from individual consumers such as factories, buildings, and residences to regions and towns.

It is impossible to discuss the situation of the Japanese energy sector without mentioning the Great East Japan Earthquake that occurred on 11 March 2011 and a serious accident at the Fukushima Daiichi Nuclear Power Plant, as both marked a major turning point in Japan's energy policy. An early post-2011 energy vision for both recovery from the earthquake and for the future was based on a shift to renewable energy and decentralised power systems that do not depend on large-scale infrastructure (National Strategy Office of Japan, 2011). This was a shocking and discontinuous experience for the Japanese energy sector based on nuclear power. However, as the shift to renewable energy has been a globally common strategy, and the trend towards decentralised energy has been a gradual process, these two directions were natural choices for Japan. Nevertheless, these were not the only means used at that time to address the urgent challenges the country faced.

Immediately after the Great East Japan Earthquake, Tokyo Electric Power Company (TEPCO) was forced to implement rolling blackouts in its supply area (Tokyo and Tohoku), which were the first scheduled blackouts in Japan since end of World War II (Sokołowski, 2022a, p. 47). Moreover, to prevent such supply disruptions from recurring, various measures related to DR to reduce electricity usage were introduced (Sokołowski, 2022a, p. 47). In 2012, the specified study committee was established by the Ministry of Economy, Trade, and Industry (METI), to determine rules and communication standards for DR for business players. Soon, demonstration tests were started at a national level. Since 2016, these activities have been extended to the framework of VPP which not only reduces power consumption but also controls various energy resources on the demand side, just like a power plant. This was accompanied by demonstration tests of the effectiveness of the system along with the introduction of requirements for the balancing market aiming at improving aggregators' systems and operations, as well as their business environment (Asano, Takano and Ishii, 2024).

Apart from the fact that such demand-side intelligence applications have been an important technological element for local production and consumption of renewable energy in Japan, since early 2010s they have been the elements of demonstration projects for building smart communities (Japan for Sustainability, 2010). Moreover, after the 2011 Great East Japan Earthquake, implementing DR became an essential element in Japanese smart communities. Furthermore, with the development of new technologies and solutions that apply AI, digitalisation has advanced in many fields, including the energy sector (Sokołowski, 2022b), leading to generation and utilisation of large volumes of digital data (Big Data).

In this context, Japan proposed a concept of "Society 5.0" (Cabinet Office, 2021). The concept relies on highly integrated cyberspace (virtual space) and physical space (real space). It is aimed at creating a human-centred society that achieves both economic development and the resolution of social issues. By fully leveraging highly sophisticated AI to connect, analyse, and transform Big Data from various fields, the core of the mission of Society 5.0 is to create services and businesses that offer unprecedented value. For these reasons Japan offers financial subsidies provided through frameworks such as the Cabinet Office's Digital Garden City Nation Initiative (Jiji Press Photo, Ltd., 2021) and the Ministry of the Environment's Decarbonization Leading Regions (Ministry of the Environment of Japan, 2022) allowing local governments and communities to capitalise on their unique characteristics. The journey towards realising Society 5.0 has only just begun.

4.3 Smart Metres and Data Usage: Current Situation in Japan

Although the advantages of employing information and communication technologies for remote and automated metre reading of consumers' electricity and gas usage have been extensively discussed, their implementation in Japan has not materialised primarily due to cost-related challenges. However, due to price reduction of electronics and their rapid penetration into Japanese market, in 2011, METI established a policy to introduce smart metres for electricity and gas with the aim of understanding consumer energy profile, changing consumers' behaviour, and providing new services that utilise energy usage information to save energy and ultimately realise a decarbonised society. Later the functions and specifications of first-generation smart metres were determined, and their installation was scheduled to be completed for about 81 million customers in 2024. Gas and water metres are also being made "smart".

With respect to details, electric power smart metres are owned by general power transmission and distribution companies and measure the electricity usage of consumers over a 30-minute period. The readings are transmitted to the head end system (HES) and metadata management system (MDMS) via the communication network (route A). General electricity transmission and distribution companies are required to send the metre readings of consumers to each retailer within one hour. The measurement granularity of 30 minutes was determined from the transaction time frame of the electricity wholesale market. This specification is crucial for implementation of DR and electricity price linked to the wholesale market. Smart metres have a data communication function of consumers (route B) to notify electricity usage in response to requests from consumer's device such as home energy management systems (HEMS). In addition to billing, retail businesses also provide services such as graphic display of power usage for time of use price plan and energy conservation consulting through the Internet.

From the perspective of promoting digital transformation in the power sector, the standard functions of next-generation smart metres have been investigated at the council and announced with the perspective of introducing it from 2025. Figure 4.1 shows an overview of the smart metre system and acquired data.

The elements added to the next-generation specifications compared with the first-generation specifications are shown in bold characters in Figure 4.1. The main additions to the next-generation specifications include measuring reactive power and voltage. It was also determined that the next-generation smart metres measure 5-minute values of active power, reactive power, and voltage, and provide them within a few days. These measurements are expected to be applied for modernising the planning and operation of the Japanese distribution grid to expand the amount of renewable energy that can be connected (hosting capacity) and reduce power transmission loss. The next-generation metres will include support for 15-minute segmentation and acquisition of 1-minute value of effective power amount via route B. Furthermore, it is required that route B would support WiFi, widely used communication media, and that an IoT route would be prepared as a communication line with other devices such as gas and water metres and storage batteries.

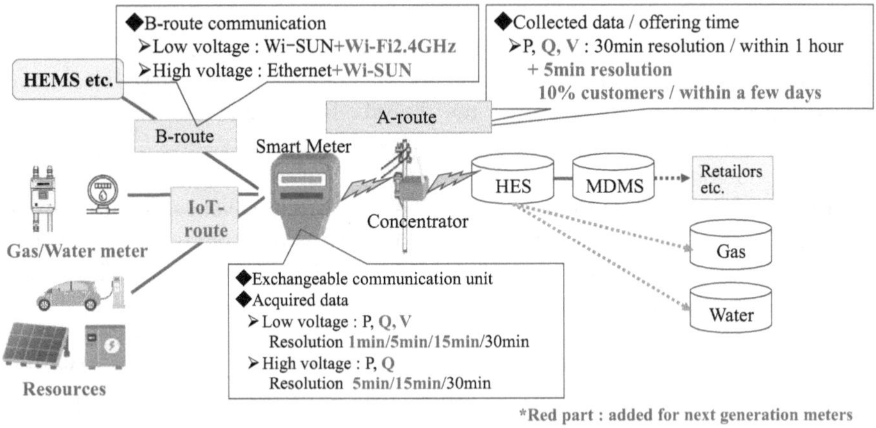

Figure 4.1 Smart meter and related system in Japan.

Based on this basic specification, as of 2024, the next generation of Japanese smart metres is currently under development.

In the realisation of next-generation smart metres, there are still some undecided aspects, such as the implementation of IoT routes and the details of how to connect gas and water metres. Therefore, while keeping a close eye on the development of next-generation smart metres, we will continue to develop the required functions for the bidirectional multimeter gateways as one way to realise the functions we pursue.

The introduction of smart metres will not only enable efficient and detailed bill settlement but also enable various uses of data, such as improving the efficiency of energy use by consumers, monitoring services for the elderly, and providing evacuation guidance for local governments in the event of a disaster. On the other hand, power consumption data also contains information such as consumers' lifestyle patterns and their presence or absence at home. Therefore, sufficient consideration must be given when using the data, including protection of personal information, due to the significant social utility value of power consumption data (Shimpo and Mikusek, 2023).

4.4 Distributed Cooperative Energy Management and Simulation Tool for Quantitative Evaluation of Smart Cities

In the analysis and management of the electricity grid, it becomes crucial to enhance the integration of renewable energy sources, particularly solar power generation. This effort aims to maximise the potential for sustainable energy adoption. In addition, in the future world where various energy resources will be introduced to consumers, and DR and VPP will be implemented based on optimal energy management for each individual consumer, it will be necessary to consider the overall optimisation perspective that coordinates with network energy management through electricity value transactions. The optimisation index primarily focuses on economic efficiency (energy costs). However, CO_2 emission reduction has been emerging as a crucial metric as we progress towards carbon neutrality. In terms of policy and investment decision-making, the quantitative comparative evaluation of these indicators before and after implementing specific measures is expected to gain greater significance.

In order to respond to these social needs, we at the Advanced Collaborative Research Organization for Smart Society (ACROSS) are implementing measures such as the large-scale introduction of photovoltaic (PV) generation and distributed energy resources. Additionally, we foster the cooperation between power grids and consumer EMS. Incorporating these elements, we calculate how a power grid that spreads across the city level will behave in response to a desired scenario. We calculate data such as power flow distribution, voltage distribution, and energy loss density. To achieve this, we have developed a digital simulation method (digital platform), which allows us to quantitatively evaluate indicators such as CO_2 emissions (Hayashi *et al.*, 2018).

Figure 4.2 outlines our approach. We establish the power grid according to the target area, utilising published data on land use and past research results on consumer power consumption patterns. This enables us to construct a realistic power system at the city level in cyberspace. Subsequently, we explore various scenarios

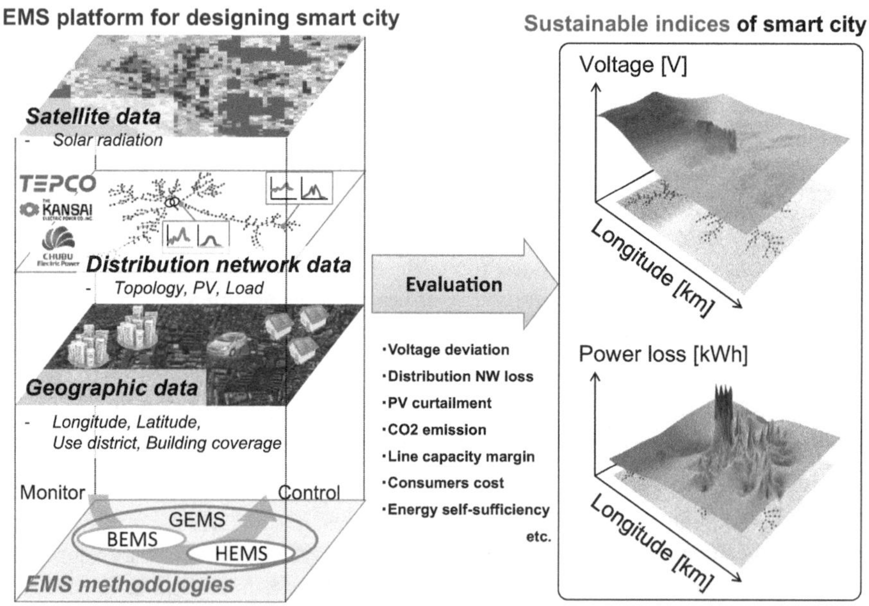

Figure 4.2 Cooperative EMS simulation platform.

related to the introduction of new equipment that contributes to decarbonisation, such as PV and electric vehicles. Specifically, we calculate the power generated by the solar power equipment using the ground solar radiation data from the meteorological satellite Himawari-8 every 2.5 minutes within a 1 km^2 section. Additionally, we consider EMSs (HEMS for households, BEMS for commercial buildings) for each consumer. These consumer-level EMSs can be shared with the grid-side EMS (GEMS) at any time resolution. On the right side of Figure 4.2, we illustrate an example of voltage distribution in the distribution grid and the power loss distribution occurring in the distribution feeders.

To accurately handle electrical phenomena such as power flow and voltage variation, one must consider various specifications of electric wires (such as thickness, length, impedance, and maximum allowable current, etc.), switch position, and the number of pole transformers. In collaboration with the Japanese energy companies, TEPCO Power Grid, Kansai Transmission and Distribution, and Chubu Electric Power Grid, we obtained the actual power flow data in 30-minute increments for days when the power load is at its maximum and minimum. This data, collected from 123 existing feeders, forms the basis for a power distribution line model. The model is now accessible to the public via the website of the ACROSS, facilitating its widespread use in research. These tools enable quantitative evaluation of the impact of equipment and technology introduction, as well as policies and investments. For instance, they allow assessment of factors such as the amount of PV systems introduced, the placement and charging patterns of electric vehicle chargers, and the availability of usable data (Fujimoto, Ishii and Hayashi, 2020).

The introduction of next-generation smart metres will be a powerful data collection source for simulation. These would be statistics based on several data points, with the spatial distribution and temporal changes of power consumption and reverse power flow from solar power generation estimated. Such data is expected to have a variety of applications, including energy management (Miyasawa *et al.*, 2021), DR potential evaluation and prediction, and estimation of solar power generation and actual load from net load (Miyasawa *et al.*, 2023), all while maintaining a focus on personal information protection.

4.5 Concept of 5Grids in Practice: Building a Multimeter Gateway

At the ACROSS of Waseda University, we are pursuing the following activities by capturing the trends mentioned in the previous section, focusing on the three D's; making maximum efforts to contribute to *Decarbonisation*, incorporating the trend of *Decentralisation*, and making full use of the fruits of *Digitalisation*. Our philosophy is highlighted by the following quote:

> [b]y collecting and integrating a wide variety of data created by people and utilising the information through AI analysis, we will contribute to improving the convenience and safety of social activities and citizen's lives. Furthermore, we will support local government in terms of policies and economic revitalization, such as optimising urban infrastructure and public transportation as

well as business innovation at companies. From these perspectives, we will start new research activities that will contribute to the realisation of the coming decarbonized, ultra-decentralized, digital society.

As a materialisation of our research activities, we are working to build a foundation for utilisation of information by referring to 5Grids. Our aim is to lead a new era of infrastructure that can respond to social needs and technological innovation, promote the use of infrastructure data, make infrastructure investments optimised, reduce social costs, improve resilience in the event of a disaster, and contribute to the safety and security of citizens as local governments and companies will strive to build next-generation smart cities focused on decarbonisation. Regardless of location and time, we believe that a mechanism for collecting data and a system for managing and utilising it are crucial elements.

Our research has a practical dimension. We have begun constructing a bidirectional multimeter gateway for residences to collect data. The basis of resident activities is data on electricity, gas, and water usage. As for electricity consumption, the system accumulates digitised data at a granularity of 15–30 minutes. In response to the trend of digital transition (DX), there is a growing interest in the introduction of smart metres and implementation of data management for gas and water. In addition a form of joint metre reading that combines with the data collection infrastructure of the preceding electricity metres is being considered in the ACROSS. Moreover, with the increasing adoption of various energy devices – such as energy storage batteries, electric vehicles (EVs), or "EcoCute" (heat pump water heaters) – within households, their role in VPPs becomes crucial. In this context, there arises a need for accurate evaluation of each device's contribution. Specifically, this evaluation should aim to measure the power input and output of each device, allowing for effective utilisation of the measured values.

In this light, in the ACROSS we are working on the development of a bidirectional multimeter gateway that aggregates and collects all the aforementioned data. This metre has a communication function and can send and receive data in both directions. The data sent from the metre to the device is used to control the device in VPP and to close the gas metre in case of an emergency. Figure 4.3 shows these functions. One can see that the multimeter gateway covers the specification of the next-generation smart metres mentioned in Section 4.3.

Another essential elemental technology is a data platform that integrates and manages all the data collected for each house. In addition to the basic function of distributing collected data to electricity, gas, and water utilities, this platform also links information with external systems to obtain weather information, various data related to traffic, data for human flow, etc. Then, it performs various learning, analyses, predictions, etc. with the collected basic infrastructure data (Figure 4.4). Furthermore, through application programming interface (API), local governments and businesses can obtain the information they need. Such a system supports customers' participation in VPP by utilising infrastructure data generated

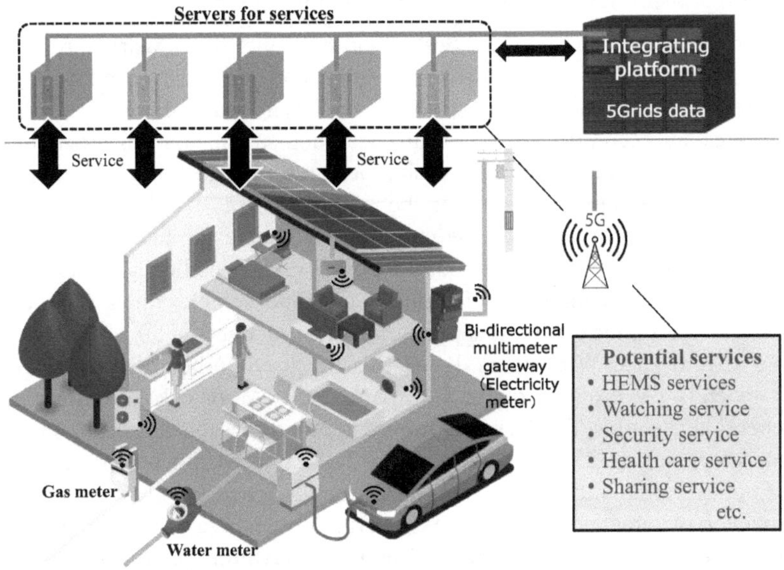

Figure 4.3 Bidirectional multimeter gateway in a house.

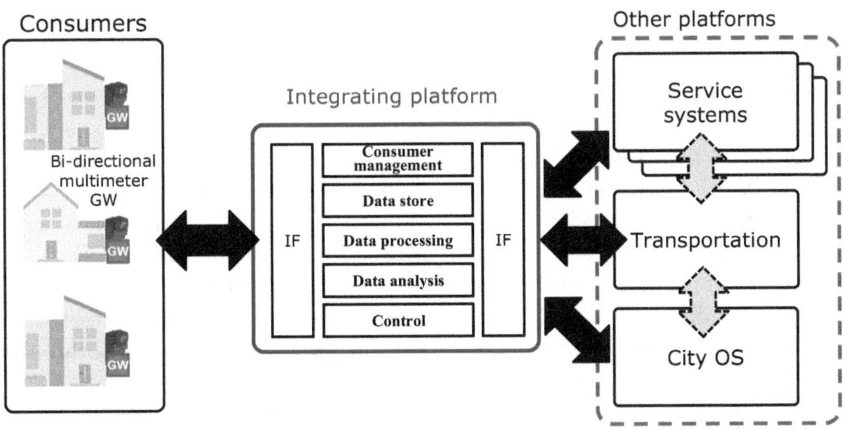

Figure 4.4 Image of data integrating platform.

by customers, and serves as the basis for improving the sophistication of municipal administrative services and creating new services.

There are various possible patterns as to which business operator will own, operate, and provide services for the aforementioned systems, and here we will conduct a technology-specific study.

4.6 Sector Coupling of Electricity and Transportation

Electricity and transportation are both basic elements of infrastructure that are essential to human life, but it can be said that until now they have rarely been considered in conjunction with each other. In recent years, the electrification of automobiles has been progressing toward decarbonisation, and it is inevitable that this will accelerate further with the global declaration of carbon neutrality, so the relationship between the two is rapidly becoming closer.

The transition from gasoline, the main energy source for automobiles to date, to electricity will significantly impact the power grid. Specifically, to achieve EV charging times comparable to refuelling a gasoline vehicle, quick chargers are essential. In Japan, the standard capacity for these chargers is 50 kW, but there is a move toward 150 kW chargers due to the increased on-board battery capacity, which extends EV cruising range. Compared to 3 kW or 6 kW contract power for general households, this charging infrastructure represents a substantial system capable of powering 25 to 50 households. The rapid establishment of charging infrastructure is imperative for energy transition.

However, the practice in Japan does not typically involve constructing or extensively reinforcing the distribution grid, and the prevailing approach is to utilise existing facilities as efficiently as possible. Notably, a change in recent years has been caused by the mass introduction of PV, and there is no doubt that this will continue to progress in the future. With the spread of PV, there arises the possibility of power surpassing consumer electricity demand, resulting in a reverse power flow injected into the distribution grid. This situation can lead to voltage levels exceeding the upper limit. In such instances, proactive countermeasures, including equipment reinforcement, become necessary.

Considering these situations faced by each sector (electricity and transportation), the merits of linking the two sectors become evident. Specifically, if electric vehicles (EVs) can be charged during periods when reverse power flow occurs from PV systems, it is anticipated that the solar-generated electricity can be efficiently utilised, thereby alleviating strain on the power grid (see Figure 4.5).

In this way, sector coupling refers to the synergistic creation of new value by interlinking different sectors and leveraging data from each domain that have traditionally been optimised separately. This concept finds application within the data collection and utilisation framework at the core of 5Grids. Our specific focus lies on the introduction of electric buses and the revitalisation of the public transportation network in the city of Utsunomiya, situated 130 km north of Tokyo. This incentive encompasses the newly constructed light rail transit (LRT) and the comprehensive re-evaluation of bus routes. The ACROSS' research endeavours centred around integration data from the electricity sector, including smart metres as discussed in Section 4.3, as well as information from the transportation sector, such as bus travel information (together with bus location) and traffic/road conditions. Additionally, we incorporate human flow data obtained from mobile

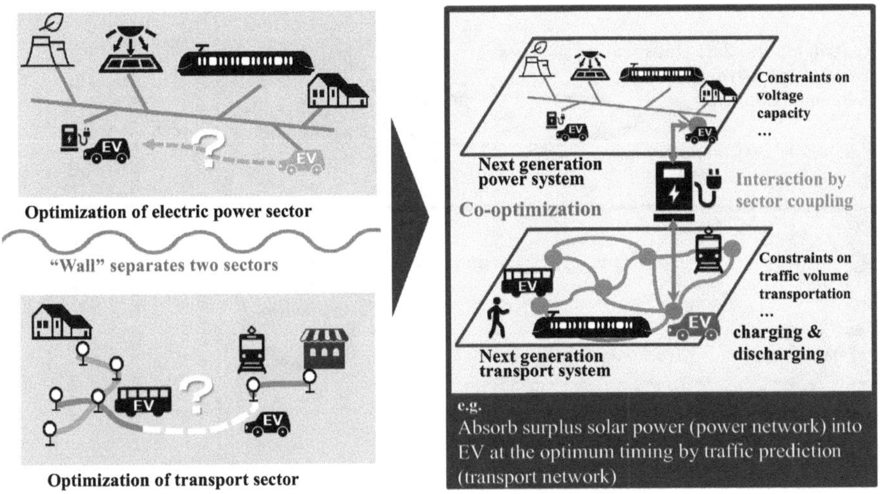

Figure 4.5 Basic idea of sector coupling between electric power and transportation.

phones. This multifaceted project is named E-MaaS (Electricity, Environment, and Mobility as a Service).

The basis for this is the digital simulation technology for city-level power systems addressed in Section 4.4. We are also actively developing a traffic simulation model and conducting research to combine this power grid. By doing so, we can dynamically evaluate energy consumption and human movement, while effectively accounting for the interaction and constraint conditions between power and transportation systems. Our primary objective is to holistically address efficient public transportation operations. Through quantitative evaluations, including assessment of CO_2 emissions and public transportation usage rates via simulations, we aim to design a transportation network centred around public transportation. This approach minimises environmental impact and reduces stress associated with movement. Our ultimate goal is to create technology that contributes to policy planning (see Figure 4.5). The integrated evaluation platform for power and transportation serves as a virtual representation of a city in cyberspace. It relies on digital simulation based on the acquisition, analysis, and prediction of big data on power, transportation, and human flow. We refer to this platform as the "spatio-temporal multi-dynamics prediction engine".

In the pursuit of optimally designing the basic infrastructure of electricity and transportation, it is essential to adopt a perspective that serves the needs of local communities from a *people-centric* standpoint. In Utsunomiya City, the ACROSS aims to support public transportation to enhance vulnerable groups such as the elderly and children in collaboration with the city government. Simultaneously, we

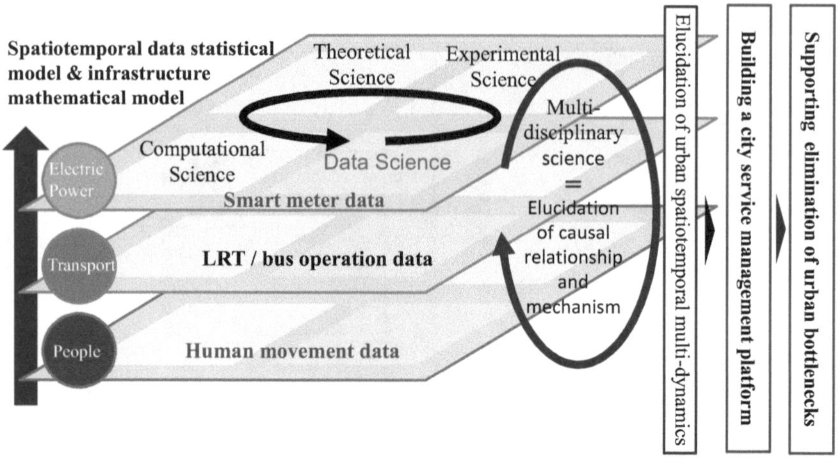

Figure 4.6 Spaciotemporal multiscale dynamics to design a city.

prioritise improving the comfort and convenience for both residents and visitors through tourism initiatives and events, fostering a vibrant urban environment.

Another important element is the perspective of ensuring human safety and security through high-level disaster-resistant resilience. EVs, beyond their role as conventional cars, possess the feature of serving as mobile storage batteries. Deploying EVs at evacuation centres and critical locations in the event of a disaster and when the power supply is disrupted enables utilisation of stored electricity. The overall goal of E-MaaS is to enrich people's living spaces while reducing CO_2 emissions and increasing public transportation share (Figure 4.6).

4.7 Conclusion

As discussed in this chapter, the energy transition in Japan requires the decentralisation and progress of demand-side technology, both fundamental for smart cities. The presented idea of 5Grids, which provides the mechanism of data collection and application, together with related technology research, contributes to developments of smart cities, also by sector coupling between power and transportation.

Moving forward, targeted smart city development will be promoted across different regions in Japan, leveraging the unique characteristics of each area. This will be done through frameworks like Decarbonization Leading Regions and the Digital Garden City Nation Initiative. Through close collaboration with local governments and industries, the ACROSS intends to generate practical use cases for the 5Grids and develop sustainable business models based on effectiveness verification and continuous enhancement of individual technological components.

Acknowledgements

This work was supported by the JST CREST (grant number JPMJCR15K5) and by Council for Science, Technology and Innovation (CSTI), Cross-ministerial Strategic Innovation Promotion Program (SIP), "Energy systems of an Internet of Energy (IoE) society" (funding agency: JST).

References

Asano, H., Takano, H. and Ishii, H. (2024) "Distributed Energy Resource Integration for Carbon Neutral Power Systems: Market-Based Approaches to Ancillary Services and Microgrid Operation", *IEEJ Transactions on Power and Energy* , 19, pp. 598–607.

Cabinet Office (Japan) (2021) "Society 5.0". Available at: www8.cao.go.jp/cstp/english/society5_0/index.html (Accessed: 11 April 2024).

Fujimoto, Y., Ishii, H. and Hayashi, Y. (2020), "Designing Sustainable Smart Cities: Cooperative Energy Management Systems and Applications", *IEEJ Transactions on Electrical and Electronic Engineering*, 15(9), pp.1256–1270.

Hayashi, Y. *et al.* (2018) "Versatile Modelling Platform for Cooperative Energy Management Systems in Smart Cities", *Proceedings of the IEEE*, 106, pp. 594–610.

Humans (2023) "The Characteristics of Japanese Culture". Available at: https://vocal.media/humans/the-characteristics-of-japanese-cultureedia (Accessed: 2 March 2024).

Ishii, H. *et al.* (2016) "Demand-side response/Home energy management", in Liu, C.C., McArthur, S. and Lee, S. (eds.), *Smart Grid Handbook*, Vol. 3. West Sussex: John Wiley & Sons, pp. 1235–1247.

Ishii, H., and Hirohashi, W. (2019) "Infrastructure system supporting VPP", *Journal of IEEJ*, 139(3), pp. 144–147.

Japan for Sustainability (2010) " 'Next-generation Energy and Social System Demonstration' Project Begins in Four Regions". Available at: www.japanfs.org/sp/ja/news/archives/news_id030368.html (Accessed: 11 April 2024).

Jiji Press Photo, Ltd. (2021) "Japan Eyes Huge Grants for 'Digital Garden City' Initiative". Available at: https://sp.m.jiji.com/english/show/15937 (Accessed: 1 November 2023).

Ministry of the Environment of Japan (2022) "Decarbonized Community Development Support Site". Available at: https://policies.env.go.jp/policy/roadmap/preceding-region/#main (Accessed: 1 April 2024).

Miyasawa, A. *et al.* (2021) "Spatial Demand Forecasting Based on Smart Meter Data for Improving Local Energy Self-Sufficiency in Smart Cities", *IET Smart Cities*, 3(3), pp. 107–120.

Miyasawa, A. *et al.* (2023) "Forecast of Area-scale Behaviors of Behind-the-meter Solar Power and Load Based on Smart-metering Net Demand Data", *IET Smart Cities*, 5(1), pp. 19–34.

National Strategy Office of Japan (2011) "Interim Compilation of Discussion Points for the Formulation of 'Innovative Strategy for Energy and the Environment'". Available at: www.cas.go.jp/jp/seisaku/npu/policy09/pdf/20110908/20110908_02_en.pdf (Accessed: 11 November 2023).

Sano, T. (2011) "Improving Efficiency of Thermal Power Generation in Japan", *Journal of Power and Energy Systems*, 5(2), pp. 146–160.

Shimpo, F. and Mikusek, P. (2023) "Personal Data Protection Issues in the Context of Smart Cities Solutions", in M.M. Sokołowski, A. Visvizi (eds) *Routledge Handbook of Energy Communities and Smart Cities*. London: Routledge, pp. 205–219.

Sokołowski, M.M. (2022a) *Energy Transition of the Electricity Sectors in the European Union and Japan: Regulatory Models and Legislative Solutions*. Cham: Palgrave Macmillan.

Sokołowski, M.M. (2022b) "Artificial Intelligence and Climate-Energy Policies of the EU and Japan", in Bielicki, D.M. (ed.), *Regulating Artificial Intelligence in Industry*. London: Routledge, pp. 138–155.

5 Sustainable Waste Management for Carbon-Neutral and Circular Economy in Japan

Jeongsoo Yu, Xiaoyue Liu, Tadao Tanabe, Gaku Manago, and Shiori Osanai

5.1 Introduction

Incineration treatment has been used by most big cities in Japan for waste management since the period of rapid economic growth. In 2000, the Basic Law for Promoting the Formation of a Recycling-Oriented Society was enacted. The enforcement of recycling laws has been based on this legislation, as it covers rules for containers and packaging including plastic materials, end-of-life vehicles, electric appliances, food, construction materials, and small household appliances. Despite this, roughly 80% of municipal solid waste (MSW) is still being burned in Japan (Ministry of the Environment, 2023).

In recent years, developed countries have focused on carbon neutrality and the circular economy, and Japan's waste administration is trying to switch from energy recovery to material recycling. For example, in 2022, the Plastic Resource Circulation Act was enacted, and all waste plastics from households were collected and actively promoted for material recycling. On the other hand, there are many dirty and complex plastics in plastic waste from each local government. Collecting a large amount of clean and identical materials is necessary to produce and supply high-quality recycled materials. Moreover, the development of a recycling network that can provide reliable and high-quality recycled materials through cooperative recycling behaviour and sorting technology is essential.

In this context, the main goal of this chapter is to provide an overview of the waste management and recycling policies in Japan. Additionally, the chapter proposes approaches and strategies for building a new smart city that controls a carbon-neutral and circular economy using a sustainable recycling system. The opening section presents an overview of Japan's waste policy and details the current situation regarding waste incineration with energy recovery. It also discusses new recycling technologies that have been introduced in Japan to utilise terahertz waves to evaluate degradation and identify different types of plastic waste. This new technology has great potential for decarbonisation and sustainable resource recycling in the future. Furthermore, in this chapter we highlight the characteristics of consumers' cooperative recycling behaviours and explore the possibility of new recycling methods to supplement waste administration collection routes. Also, we

DOI: 10.4324/9781003471417-5

propose an idea of constructing a sustainable recycling network in partnership with industry, academia, and government. We believe this will be a long-term project aimed at achieving sustainable urban development based on a carbon-neutral and circular economy.

5.2 Waste Management and Recycling Policy in Japan

Japan has been developing waste management and recycling policies for decades in response to environmental challenges and resource conservation goals. After World War II, Japan experienced rapid industrialisation and economic growth, which led to a consumer society characterised by mass production, distribution, and consumption. As a result, there has been an increase in waste, which is causing environmental problems. In big cities, lifestyle changes have contributed to the diversification and complexity of waste composition. Due to Japan's small land area and high land prices near urban areas, the selection and securing of final disposal sites has become extremely challenging. Local governments responsible for disposing of household waste have selected incineration as their primary method for sanitary treatment and waste reduction. Until the 1970s, the post-war waste management in Japan was mainly focused on disposal rather than recycling. After experiencing a surge in environmental problems, Japan began to pay special attention to environmental issues in the late 1960s, leading to the introduction of basic environmental regulations, soon revised and strengthened during a landmark 1970 session of the Japanese parliament named the Pollution Diet (Sokołowski and Kurokawa, 2022). In the 1980s, plastic products became more common and as the use of plastic containers and packaging expanded, many of these products were thrown away. Incineration was the main method of waste treatment in Japan. Its incineration rate was 80%, a level much higher than that of other OECD countries (Environmental Protection Bureau, City of Kawasaki, n.d.). Many plastics produce exhaust gases and ash dust during the incineration process, leading to air pollution, deterioration of water quality, and severe pollution problems. In addition, the shortage of waste treatment and disposal facilities has led to an increased interest in waste management and recycling by local governments. Waste management policies were developed with the aim of improving public health and preserving the living environment.

The Japanese government enacted the Containers and Packaging Recycling Law[1] in 1995 and started recycling programmes for glass bottles and PET bottles in 1997 to promote effective recycling. The recycling law added paper and plastic containers and packaging, and first introduced extended producer responsibility (EPR) in 2000. This law changes the traditional waste management system in which only municipalities are responsible for household waste. It is a requirement for stakeholders to share responsibility and work together to recycle containers and packaging (Ministry of Economy, Trade and Industry, 2003). As a result, consumers must separate containers and packaging waste, and municipalities provide collection and immediate treatment as a public service, and manufacturers and businesses pay for the recycling cost (Yamakawa, 2016).

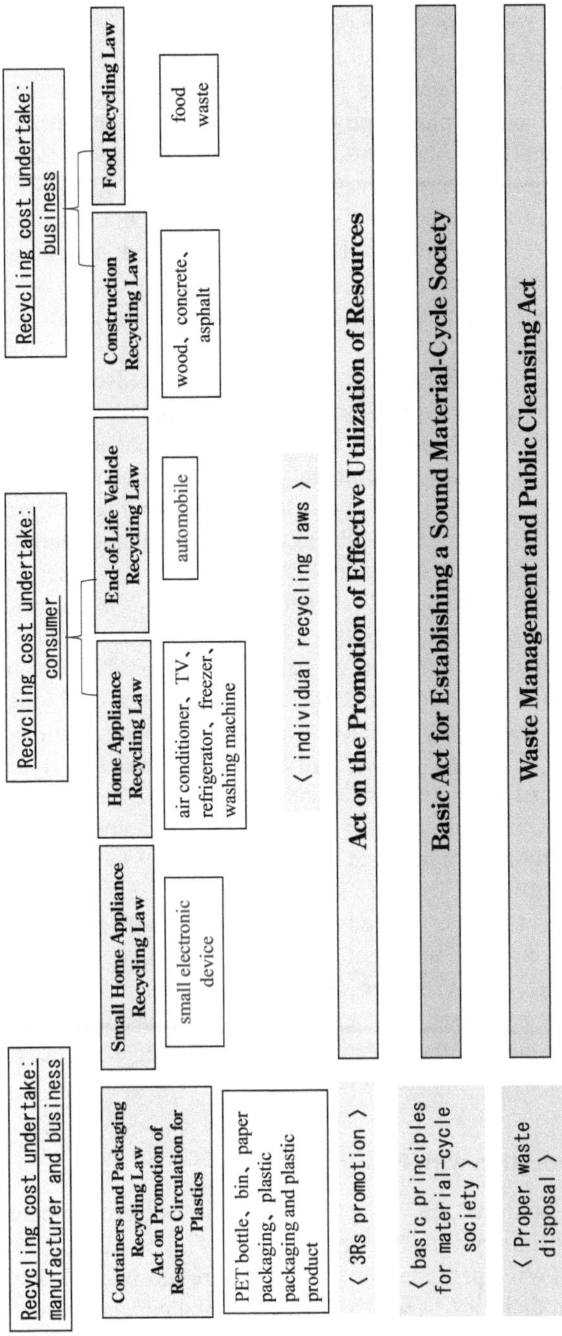

Figure 5.1 Basic laws and individual recycling laws for sustainable waste management in Japan.

In 2001, the Basic Law for Establishing the Recycling-based Society[2] was enforced. Furthermore, to promote the 3Rs (reduce, reuse, recycle), the Act on the Promotion of Effective Utilization of Resources[3] was introduced (Food and Agriculture Organization of the United Nations, 1991). As part of Japan's "3Rs+ Renewable" initiatives to promote resource circulation of plastics in each stage of the entire lifecycle of plastic products, in collaboration with all stakeholders including municipalities, businesses, and consumers, the Act on Promotion of Resource Circulation for Plastics[4] was brought into effect on 1 April 2022. In order to promote the sorting collection of plastic resources, not only waste containers and packaging, but also plastic products can be recycled. And manufacturers and vendors can prepare plans for voluntary collection and recycling of products, in cooperation with municipalities and recycling businesses (Cabinet Office, Government of Japan, 2022).

In 1998, Japan introduced the Home Appliance Recycling Law,[5] which mandated manufacturers to collect and recycle certain household appliances, such as refrigerators and air conditioners (Panasonic Eco Technology Center, PETEC, 2024). In the year 2000, the government enacted the Food Recycling Law[6] with the objective of securing effective utilisation of resources related to food and suppressing the discharge of food-related waste. On the other hand, the generation of construction waste, such as concrete blocks and wood, generated during construction projects, has been prone to illegal dumping. In 2000, Japan introduced the Construction Materials Recycling Law,[7] which requires construction material waste to be recycled in construction projects (Tokumoto, 2003). The End-of-Life Vehicle Recycling Law[8] was introduced in 2002. It required automobile manufacturers to take responsibility for recycling end-of-life vehicles (Japan Automobile Recycling Promotion Center, 2024). Also, the Small Home Appliance Recycling Law[9] started in 2013 (Ministry of Economy, Trade and Industry, 2022). Its aim was to recover rare and precious materials that are present in small home appliances like PCs, mobile phones, digital cameras, and game devices, and to encourage their recycling. Research published by the Japanese government's National Institute of Materials Science in 2010 estimated that there were 6,800 tonnes of gold recoverable from used electronic equipment in Japan. It is the concept of urban mining by Tohoku University, Japan (Yu *et al.*, 2011).

The framework for waste treatment and recycling has been developed through changes to fundamental laws and the enactment of specific recycling laws (Figure 5.1).

5.3 Waste Incineration with Energy Recovery

In 1900, Japan established the Waste Cleaning Act[10] as its fundamental waste management policy (Ministry of the Environment, 2014). The reasons for implementing waste-to-energy (WtE) practices are diverse. There has been an increase in large-scale incineration facilities, as well as environmental concerns such as dioxins and global warming. The initial catalyst for this shift was the oil crisis of the 1970s, which highlighted the need for alternative energy sources as a substitute for oil.

Since the 1990s, global environmental issues such as climate change, ocean pollution, natural resource leakage, and major disasters have increased. A concerted effort has been made to encourage the use of new energy sources, especially as a solution to global warming. One of the new energies being promoted is energy recovery from waste, which contributes to both waste reduction and energy generation. Household waste consists of organic and plastic materials, as well as paper materials. Due to the historical focus on incineration-centred waste management policies, WtE remains a viable option in the shift towards reducing fossil fuel usage. Despite the advantages, waste incineration has been criticised due to air pollution concerns, the potential release of harmful substances, and opposition from local communities (Malinauskaite *et al.*, 2017). It is still a challenge to balance the benefits of energy recovery with environmental and health considerations.

Japan has invested in WtE facilities as part of its waste management infrastructure. MSW is burned by these facilities to generate electricity and heat. Japan's WtE facilities are designed to maximise energy recovery efficiency. Waste is combusted at high temperatures by high-tech incinerators, which reduce the volume of waste and extract energy in the form of heat. However, following the principles of the Waste Cleaning Act, even relatively small municipalities in Japan have historically developed small-scale incineration facilities to carry out waste incineration. Consequently, as waste disposal is managed at the municipal level, the average generation capacity of WtE plants in Japan remains low (Table 5.1) (Ministry of the Environment, 2023). Furthermore, there are still many incineration facilities that are not able to efficiently utilise the excess heat generated. On the other hand, the heterogeneous nature of MSW presents challenges for efficient incineration. The composition of waste can have an impact on the combustion process and energy recovery quality. Each recycling policy promotes the collection and reprocessing of recyclable materials such as plastics and paper for recycling and reduces waste from incineration in waste management. In particular, the decrease in the incineration of plastics, which have a high calorific value, has had a significant impact on the generation of WtE.

Japan's greenhouse gas emissions in 2020 were 1.15 billion tonnes, which has been gradually decreasing. However, 84% of CO_2 emissions originate from energy production, such as electricity power generation, making the reduction of emissions in this sector an ongoing challenge (Eri and Akane, 2018; Ministry of the Environment, 2022). In Japan, incineration with energy recovery, which refers to "thermal recycling", involves incineration of non-combustible waste such as

Table 5.1 The status of electricity generation at waste incineration facilities in Japan in 2021

Number of electricity generation facilities	396
Total electricity generation capacity	2,149 MW
Average electricity generation efficiency	14.22%
Total amount of electricity generated	10,452 GWh

Source: Ministry of the Environment (2023).

plastics to recover heat. Thermal recycling is a unique concept in Japan and is not recognised as recycling or renewable energy in other countries. Thermal recycling is commonly referred to as energy recovery internationally, but it is considered heat recovery, not recycling.

Although waste incineration can recover energy, waste management often prioritises waste reduction over resource recycling. Recycling efforts to produce valuable materials from the waste stream may be limited by putting more emphasis on energy production. Japan's waste management policy is centred on incineration and energy recovery, which could potentially overlook the principles of priority waste management.

5.4 Challenges for Sustainable Resource Recycling

Addressing the challenges of sustainable resource recycling is important. This consists of managing and recycling waste through various strategies. The goal is to improve recycling efficiency as well as motivate cooperative recycling behaviour. Furthermore, improving environmental awareness is a long-term approach. These initiatives can tackle the problems associated with resource recycling while also fostering a long-term commitment to environmental sustainability for both individuals and societies.

5.4.1 Priority on Recycling Plastic Waste

Plastics are derived from naphtha produced in the petrochemical refining process, and their production and consumption are increasing even in developing countries due to their affordability and ease of moulding. Recently, there are many complex and diverse plastic issues to achieve decarbonisation and a circular economy, including ocean pollution, micro-plastic problems, or increasing single-use plastics, as well as China's ban on plastic waste importation. There is an urgent need for plastic recycling and the promotion of bioplastics, as the importance of the circular economy is being emphasised internationally. To achieve a high level of material recycling, it is important to efficiently sort the types and materials of waste plastic products.

The status of plastic recycling in Japan can be divided into three categories: material recycling, thermal recycling (energy recovery), and chemical recycling. Material recycling is a way to reduce CO_2 emissions by plastic recycling without damaging its material features. The process of thermal recycling involves recovering heat by burning waste plastics that have degraded so much that there is no other choice but to burn them. Chemical recycling involves decomposing waste plastics and recycling them as chemical raw materials, but energy is needed to convert polymers back to monomers (Plastic Waste Management Institute, 2022). To reduce CO_2 emissions, it is necessary to carry out degradation screening of waste plastics that have been burned in thermal recycling. We should reduce the amount of plastic waste to be burned to achieve carbon neutrality.

There are many different types of olefin-based plastics such as polyethylene (PE) and polypropylene (PP). Depending on the differences in structure and additives, these are divided into specific and a broad range of categories. Due to their widespread use in single-use packaging containers, these plastics should be considered recyclable materials, and material recycling should be the highest priority. Currently, plastic is sorted using hand, specific gravity, and near-infrared methods. However, the separation accuracy of plastic materials by existing sorters is not high as well as limited. Degradation evaluation cannot be done in principle in all these methods, which is particularly important for material recycling. Degradation occurs when molecular chains are broken, and oxygen atoms are added to them. And the vibrations between molecular chains have a significant impact on degradation. Therefore, the evaluation of degradation based on the measurement of the dielectric constant in the terahertz frequency region, which corresponds to the vibrational frequency between molecular chains, is expected to be possible.

Millimetre waves and microwaves are part of the low frequency range of terahertz. Terahertz waves can be used to characterise the properties of polymer molecules because they are close to the vibrational frequency of the molecular chains that make up polymers (Tanabe *et al.*, 2020). Terahertz waves can penetrate both plastics and dielectrics with low polarity, such as clothing, and can be highly absorbed by water, which is a polar liquid. Plastic recycling can be achieved using terahertz waves due to their promising characteristics. Terahertz waves are in the intermediate zone between radio and light waves with a frequency range of approximately 0.1–10 THz (wavelength of 30 μm to 3 mm). Terahertz waves have an energy of 1/100,000 of X-rays, and there is no risk of damaging cells or body fluids, so exposure to the human body is safe (Yu *et al.*, 2022).

Our research team is examining the terahertz characteristics of polymers. For example, the absorption peak can be observed shifting at a high frequency due to the low-density PE samples which are degraded by ultraviolet (UV) radiation translational vibrations around 2.1 THz (Iwasaki, Kanehara, and Tanabe, 2023). The higher frequency shift is caused by the shortening of molecular chains and their increased mobility during UV degradation. Figure 5.2 illustrates how the peak shift and breaking strain are linked in a mechanical tensile test. The breaking strain decreases with UV degradation, which is caused by the short-circuiting of molecular chains. This result indicates the possibility that terahertz waves can detect UV degradation of PE, which provides an additional value to conventional evaluation methods for polymer degradation.

Creating a smart city requires resource recycling within the city where plastic waste is generated. Plastic food containers, packaging, and waste papers are collected by several companies in Japan, but they focus on limited plastic waste such as foam trays, egg cartons, and PET bottles. Predefined materials make it possible to collect these items as a single material. However, transparent containers used for food packaging, made of materials like polyethylene terephthalate or

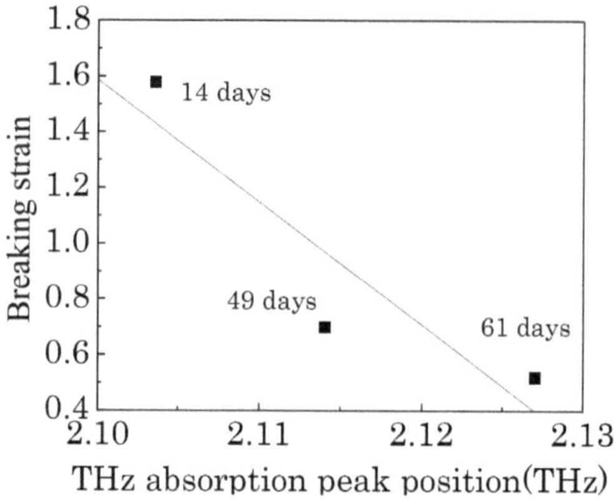

Figure 5.2 Relationship between absorption peak at 2.1 THz and breaking strain of LDPE under UV irradiation.

Source: Author's own elaboration/research.

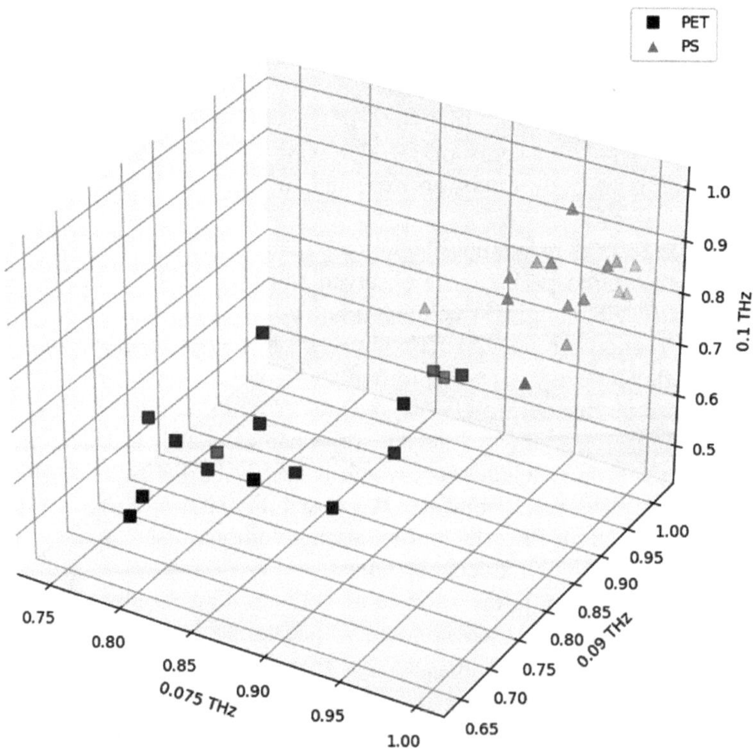

Figure 5.3 Identification of transparent plastic materials by using three light sources.

Source: Author's own elaboration/research.

polyethylene, pose difficulties in sorting due to their indistinct appearance, leading to collection of many non-recyclable materials (Iwasaki , Kanehara, and Tanabe, 2023; Iwasaki et al., 2024).

Identifying the material of transparent containers is necessary to recycle this plastic waste. As a result of changing lifestyles, many companies no longer have internal cafeterias, and most people prefer to purchase food in transparent containers from convenience stores and supermarkets. For a circular economy, voluntary collection of transparent containers and material identification are crucial.

Plastic waste material identification is being conducted using terahertz waves. PET and PS materials were identified by using three light sources at frequencies of 0.075, 0.09, and 0.1 THz. While many use broad frequency bands, such as near-infrared, for material identification, it is known to be challenging for transparent plastics. As shown in Figure 5.3, utilising terahertz waves makes it possible to classify polyethylene terephthalate and polystyrene containers, even transparent materials.

5.4.2 Cooperative Recycling Behaviour of Waste Paper

Every year, Japan generates a large amount of waste paper. The collection rate of waste paper in Japan is at a high level. However, the high rate of waste paper collection is dependent on the collective efforts of community groups. Local communities in Japan are vital in recycling because they have neighbourhood associations that handle waste collection and sorting. These groups make sure that waste paper and other recyclables are properly separated and processed, which leads to a sense of community responsibility towards recycling.

The Containers and Packaging Recycling Law mandates the recycling of waste paper and recommends that municipalities provide waste paper collection as a public service. Therefore, municipal authorities are responsible for the transportation and collection of waste paper from households, which is the primary recycling route. This system highlights the institutional framework established for recycling processes at the local government level. Each municipality sets its own collection schedule and rules, which are strictly adhered to by residents. Households are required to separate their waste into categories, including burnable, non-burnable, and recyclable materials. Waste paper is typically classified as recyclable. Residents have the responsibility of separating their waste paper from other types of waste. All types of waste paper, such as newspapers, magazines, cardboard, and other paper packaging, are collected. However, due to the infrequent kerbside collection schedules of waste recycling administrations, residents are often required to store waste paper at home for at least a week. Additionally, residents may find the municipal paper recycling service inconvenient due to the restrictions on specific days and times for disposal. Consequently, some residents opt not to separate their waste, leading to instances where recyclable paper wastes are discarded as burnable waste.

Several recycling companies have actively participated in recycling initiatives recently. They adopt policies that encourage recycling and sustainable practices, both within their operations and in the broader community. Some recycling companies have begun setting up their own waste paper collection stations in order to create their own waste paper collection system. In some recycling companies, they are also forming partnerships with supermarkets by setting up waste paper collection stations at major supermarkets (Figure 5.4) (Liu *et al.*, 2021). The collection station will provide consumers with points, which can be used in the supermarket, based on the weight of their disposal. Residents can conveniently and quickly dispose of newspapers they've finished reading by using waste paper collection stations, eliminating the need to store them at home for extended periods (Liu *et al.*, 2021). This is particularly convenient for nearby residents, who can even bring their waste paper on bicycles (Figure 5.5). This accessibility increases the efficiency and ease of recycling.

Figure 5.4 Waste paper collection stations at supermarkets in Japan.

Figure 5.5 Bicycle users transporting waste paper.

The roles of each stakeholder in the waste paper recycling system are shown in Figure 5.6. The success of Japan's waste paper collection system is largely attributed to the high level of public awareness and participation, municipal collection services, and the involvement of private recycling businesses. In recent years, while unclean plastic containers and packaging have been increasing from separated plastic waste, there has been a slight decrease in the mixing of paper materials with burnable waste. This trend shows a growing awareness and effectiveness in cooperative recycling behaviours, even though there are still challenges in managing plastic waste.

5.4.3 Promoting Environmental Awareness through Social Engagement

To improve environmental awareness, it is important to provide opportunities for everyone to understand environmental issues that are familiar to them and to

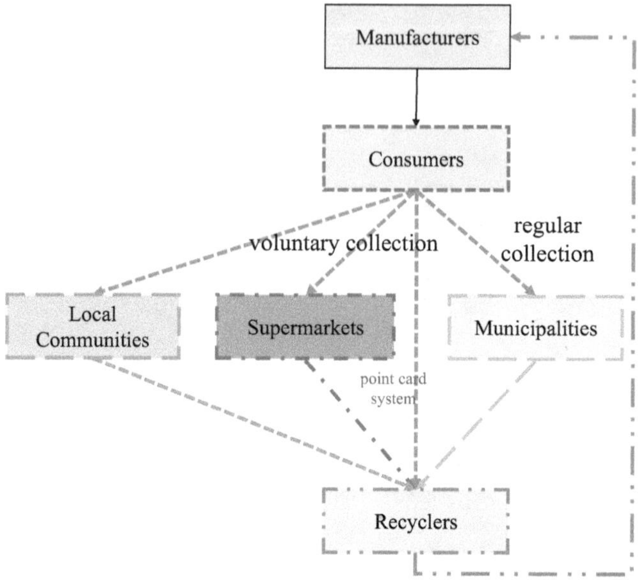

Figure 5.6 Stakeholders in the waste paper recycling system in Japan.

Source: Author's own elaboration/research.

consider some solutions to these issues. There are various opportunities available, but education plays a crucial role in changing individuals' mindsets to be more environmentally friendly. Environmental education is a type of education that improves environmental awareness. In the 1950s, Japan experienced a significant increase in economic growth, leading to pollution problems with the infamous "four big pollution diseases" (Sokołowski and Kurokawa, 2022). For this reason, pollution education and nature conservation education began to be conducted as part of environmental education, as public opinion demanded solutions to environmental problems such as pollution.

In Japan, environmental education is conducted in school. A typical example of environmental education provided in school education is "tours of waste treatment facilities". The survey results that reveal the popularity of waste treatment facility tours as a form of environmental education are described later. Miyoshi conducted a questionnaire survey among 276 teachers at 8 waste treatment facilities in Osaka prefecture (Yuji *et al.*, 2014). The results revealed that 59.1% of the teaching materials used by teachers regarding waste learning were supplementary readers, and the supplementary readers included introductions to local waste treatment facilities, such as incineration

Figure 5.7 Lecture at Naruse Oka Elementary School.

plants. Furthermore, the satisfaction level of teachers with the tour of the waste treatment facility was high at 4.4 out of 5, indicating that teachers felt that the waste treatment facility was easy to use as an environmental education facility. In this way, collaboration between schools and non-school facilities such as MSW treatment or recycling facilities can create opportunities for students to learn more authentic information.

However, even though these facilities can provide productive information to students, there is very limited knowledge to understand environmental issues. To address this issue, there are also examples of environmental education conducted through collaboration and cooperation as part of social contribution. The Yu's Laboratory at the Graduate School of International Cultural Studies, Tohoku University, has been continuously providing recovery support education in Higashi Matsushima city since the Great East Japan Earthquake. As part of a reconstruction support education project, the Yu's Laboratory visits elementary schools in the disaster-stricken areas and provides environmental education on resource recycling through delivery classes (Figure 5.7).

Higashi Matsushima City Naruse Oka Elementary School

Case Study

On 28 June 2022, the Yu's Laboratory held a visiting lecture at Higashi Matsushima City Naruse Oka Elementary School as part of its social contribution. The lecture was led by Professor Yu, who explained plastic waste issues and recycling in the context of the SDGs. This class was conducted through hands-on learning in which students directly interacted with garbage from developing countries, recycled materials, used automobile parts, and old electric appliances such as cellular phones, digital cameras, and game machines. And students tried to use a plastic waste sorter that was developed by Professor Yu's research group. In the SDGs lecture by Professor Yu, he explained the current state of waste management issues in each country, recycling methods, and the impact recycling has on local people. In the hands-on learning session, the students were able to come into direct contact with waste plastics from various countries and learn about various types of recycling through dialogue with Professor Yu and students from his laboratory. Students also learned that some recycling in developing countries is done entirely by hand, while others use cutting-edge technology, such as material identification using terahertz waves. Furthermore, there were eight stakeholders involved in this lecture, including three universities, one research centre run by the local government, and four private companies. As a result, the students realised once again the significance of starting with what they have and working on it in whatever manner they are able. And they could understand that there are many ways to recycle based on the knowledge gained from joining various contributors. Moreover, 24 out of 44 students (55%) started thinking about consciously separating trash in their future recycling activities.

Lecture at Naruse Oka Elementary School

The cases like lectures at the Higashi Matsushima City Naruse Oka Elementary School show that environmental awareness may be improved through social contribution. There are many ways to solve environmental problems, but what they all have in common is that they can be solved by people. Since mindset is directly linked to behaviour change, environmental education is important in that it can develop human resources who will be responsible for solving environmental problems. That is why it is important that schools and specialised institutions, such as universities and industries, work together to provide more authentic classes to students. In addition, collaborative education can provide students with comprehensive information on materials, research results, and specialised knowledge that cannot be accessed at school, which reduces the burden on teachers in preparing classes (Okubo *et al.*, 2021). Human resource development that is conducted in a manner that is tailored to the needs and current situation of the educational field is the ideal form of environmental education as a social engagement.

5.5 Conclusion and Recommendation

The Japanese government has declared its goal of becoming carbon neutral by 2050 (Ministry of Economy, Trade and Industry of Japan, 2022). To achieve this goal, it is necessary not only to conserve and enhance the absorption effect of CO_2 but also to reduce CO_2 emissions. To achieve carbon neutrality, it is important to reduce the consumption of fossil fuels. For example, the production of plastic should be stopped, and plastic waste should also be recycled by replacing alternatives. Waste management by using incineration systems is no longer useful in urban areas if all plastic waste is recycled.

If a smart city is constructed in the near future, it will be managed using a lot of information. Until this point, waste management and recycling systems have not been controlled based on reliable and accurate data analysis. Every city is producing a huge amount of MSW that consists of complex and diverse materials. Therefore, MSW treatment is one of the most important public services in urban government administration.

An optimised waste management strategy is necessary for a smart city. It includes waste collection, transportation, separation rules, and the main recycling process to reduce CO_2 emissions and produce recyclable materials as much as possible. For example, it is important to structure the database on MSW, including the volume and composition of waste, transportation system, location of recycling stations, waste management, recycling facilities, and citizen's behaviour. Analysing collected data is necessary to optimise a sustainable waste management system using AI technology. Also, it would be good information to construct a new smart city. These activities need to be coordinated between each stakeholder and to have a strong relationship. In this context, we offer three main recommendations for building a smart city that is a carbon-neutral and circular economy.

First, waste management policies should be changed to create a sustainable recycling system. The Japanese government is reconsidering the priority of recycling to achieve the goal of a carbon-neutral and circular economy. In other words, material recycling is a key factor in recycling methods. To optimise the recycling process, each local government must analyse the characteristics of waste composition and volume and understand the cooperative recycling behaviour of residents in material recycling. If a smart city could control necessary information and credible data, it is possible to utilise the potentials of both the public and private sectors with a strong relationship based on trust and sharing accurate information. If every waste collection and recycling station could collect related data using wi-fi and a camera, waste administration could use this information to decide on separation rules, policies, and even tax systems. And it is possible to promote cooperative recycling behaviour among citizens.

Second task is to develop advanced recycling technology. Our research team has proposed a new technology to separate plastic waste. Existing recycling technology has some limitations in clarifying the type of plastic materials and their characteristics, such as degradation and transparency. It is very important to identify the required conditions for producing high-quality recycled plastic materials from related industries. A successful recycling system that can produce high and consistent quality of recycled plastic needs to separate each plastic material and improper goods. To complete this technology, an AI system is necessary to use machine learning algorithms with big data (Yu *et al.*, 2022). In addition, a smart city should provide many types of information to develop a new recycling technology.

Finally, SDGs education has a beneficial effect on building sustainable urban development from a long-term perspective. Taking everyday issues as themes, SDG education could have a positive impact on children's consciousness and involve stakeholders. As a result, various players have established a strong network to promote sustainable recycling and social engagement. Furthermore, students can gain understanding of global and local issues through these activities. Eventually, they may give serious consideration to solving depopulation and declining birth rates through this delivery lecture. In Japan, there is an urgent need for regional revitalisation. Most small local cities, especially in the Tohoku area, face the same challenges in constructing a sustainable city (Osanai and Yu, 2023). Since smart cities are useful options with integrated sustainable waste management based on carbon-neutral and circular economy concepts.

Notes

1 In Japanese: 容器包装に係る分別収集及び再商品化の促進等に関する法律 [yōki hōsō ni kakaru funbetsushūshū oyobi saishōhinka no sokushin-tō ni kansuru hōritsu].
2 In Japanese: 循環型社会形成推進基本法 [junkan-gata shakai keisei suishin kihonhō].
3 In Japanese: 資源の有効な利用の促進に関する法律 [shigen no yūkō na riyō no sokushin ni kansuru hōritsu].
4 In Japanese: プラスチックに係る資源循環の促進等に関する法律 [purasuchikku ni kakawaru shigen junkan no sokushin-tō ni kansuru hōritsu].
5 In Japanese: 特定家庭用機器再商品化法 [tokutei kateiyō kiki saishōhinka hō].
6 In Japanese: 食品循環資源の再生利用等の促進に関する法律 [shokuhin junkan shigen no saisei riyō-tō no sokushin ni kansuru hōritsu].
7 In Japanese: 建設工事に係る資材の再資源化等に関する法律 [kensetsu kōji ni kakawaru shigen no saishigen-ka-tō ni kansuru hōritsu].
8 In Japanese: 使用済自動車の再資源化等に関する法律 [shiyō-zumi jidōsha no saishigen-ka-tō ni kansuru hōritsu].
9 In Japanese: 使用済小型電子機器等の再資源化の促進に関する法律 [shiyō-zumi kogata denshi kiki-tō no saishigen-ka no sokushin ni kansuru hōritsu].
10 In Japanese: 廃棄物の処理及び清掃に関する法律 [haikibutsu no shori oyobi seisō ni kansuru hōritsu].

References

Cabinet Office, Government of Japan (2022) "Concerning the Act on Promotion of Resource Circulation for Plastics". Available at: www.gov-online.go.jp/eng/publicity/book/hlj/html/202205/202205_09_en.html (Accessed: 14 January 2024).

Environmental Protection Bureau, City of Kawasaki (n.d.) "Japan's Experiences in Waste Management Policy-Japan's National Policy and Kawasaki City's Efforts".

Eri, S. and Akane, O. (2018) "Why Japan Finds Coal Hard to Quit". Available at: https://asia.nikkei.com/Spotlight/The-Big-Story/Why-Japan-finds-coal-hard-to-quit (Accessed: 14 January 2024).

Food and Agriculture Organization of the United Nations (1991) "Act on the Promotion of Effective Utilization of Resources". Available at: www.fao.org/faolex/results/details/en/c/LEX-FAOC099050/ (Accessed: 14 January 2024).

Iwasaki, F., Kanehara, I. and Tanabe, T. (2023) "Material Identification and Thermal Degradation Diagnosis of Plastics Using Terahertz Waves", *Inspection Engineering*, 28, pp. 1–7.

Iwasaki, K. *et al.* (2024) "Non-Contact Terahertz Evaluation of the Melting Point for PET Bottles", *Thermochimica Acta*, 736, pp. 179732 1-6.

Japan Automobile Recycling Promotion Center (2024) "The Authorized Automotive Recycling Coordinator". Available at: www.jarc.or.jp/en/recycling/ (Accessed: 14 January 2024).

Liu, X. *et al.* (2021) "Case Study on the Efficiency of Recycling Companies' Waste Paper Collection Stations in Japan", *Sustainability*, 13(20), p. 11536.

Malinauskaite, J. *et al.* (2017) "Municipal Solid Waste Management and Waste-To-Energy in the Context of a Circular Economy and Energy Recycling in Europe", *Energy*, 141, pp. 2013–2044.

Ministry of Economy, Trade and Industry (2003) "The Containers and Packaging Recycling Law". Available at: www.jcpra.or.jp/Portals/0/resource/association/pamph/pdf/law2003_eng.pdf (Accessed: 14 January 2024).

Ministry of Economy, Trade and Industry (2022) "Implementation Status of Recycling under the Act on Promotion of Recycling of Small Waste Electrical and Electronic Equipment (FY2020 Results)". Available at: www.meti.go.jp/english/press/2022/0823_003.html (Accessed: 14 January 2024).

Ministry of Economy, Trade and Industry of Japan (2022) "Green Growth Strategy Through Achieving Carbon Neutrality in 2050". Available at: www.meti.go.jp/english/policy/energy_environment/global_warming/ggs2050/index.html (Accessed: 14 January 2024).

Ministry of the Environment (2014) "The History and Current State of Waste Management in Japan".

Ministry of the Environment (2022) "Japan's National Greenhouse Gas Emissions in Fiscal Year 2020".

Ministry of the Environment (2023) "Regarding the Disposal and Treatment Status of Household Waste in Fiscal Year 2021".

Okubo, K. *et al.* (2021) "Present Issues and Efforts to Integrate Sustainable Development Goals in a Local Senior High School in Japan: A Case Studc", *Journal of Urban Management*, 10(1), pp. 57–68.

Osanai, S. and Yu, J. (2023) "Comparative Study on National Policies and Educational Approaches Toward Regional Revitalization in Japan and South Korea: Aiming to Achieve the Sustainable Development Goals", *Societies*, 13(9), p. 210.

Panasonic Eco Technology Center (PETEC) (2024) "Japanese Home Appliance Recycling Law". Available at: https://panasonic.net/eco/petec/recycle/ (Accessed: 14 January 2024).

Plastic Waste Management Institute (2022) "An Introduction to Plastic Recycling". Available at: www.pwmi.or.jp/ei/plastic_recycling_2022.pdf (Accessed: 14 January 2024).

Sokołowski, M.M. and Kurokawa, S. (2022) "Energy justice in Japan's energy transition: pillars of just 2050 carbon neutrality", *The Journal of World Energy Law & Business,* 15(3), pp 183–192.

Tanabe, T. *et al.* (2020) "Terahertz Detection of Halogen Additive-Containing Plastics", *Optics and Photonics Journal*, 10, pp. 265–272.

Tokumoto, S. (2003) "Construction Materials Recycling Law", *Japanese Journal of Real Estate Sciences*, 17(1), pp. 12–20.

Yamakawa, H. (2016) "The EPR for Packaging Waste in Japan", in *Extended Producer Responsibility-Updated Guidance for Efficient Waste Management.* Paris: OECD, pp. 269–276. www.oecd-ilibrary.org/the-epr-for-packaging-waste-in-japan_5jlr71rw5n32.pdf?itemId=%2Fcontent%2Fcomponent%2F9789264256385-18-en&mimeType=pdf

Yu, J. *et al.* (2011) "Emerging Issues on Urban Mining in Automobile Recycling: Outlook on Resource Recycling in East Asia", in Sunil Kumar (ed), *Integrated Waste Management – Volume II.* London: IntechOpen.

Yu, J. *et al.* (2022) "New Terahertz Wave Sorting Technology to Improve Plastic Containers and Packaging Waste Recycling in Japan", *Recycling*, 7(5), p. 66.

Yuji, M. *et al.* (2014) "Investigation of the Learning Effects of Elementary School Students' Tours of Waste Treatment Facilities", in *25th Research Presentation of the Japan Society of Material Cycles and Waste Management*, p. 39.

6 E-Methane and Its Future in Japan

City Gas and Smart Cities

Shinichi Kusanagi

6.1 Introduction

When considering the pillars for creating a society with virtually zero greenhouse gas emissions, a "decarbonised society", the most promising technology for decarbonising city gas seems to be methanation. In this process hydrogen (H_2) and carbon dioxide (CO_2) react to synthesise methane (CH_4), the main component of natural gas (see Figure 6.1). The decarbonised society has been a driving force for many young Japanese, particularly those living in cities. This cohort has actively advocated for the adoption of renewable energy sources like solar and wind power supported by such means as the Feed-in Tariff (FIT) and Feed-in Premium (FIP), or electric vehicles (EVs). Their enthusiasm that surrounds the perceived "eco-friendliness" of these measures is crucial to enhance eventual transition towards a fully renewable energy sector and electric transportation. Passive engagement is insufficient; young individuals must actively demonstrate their commitment to tackling climate change by embracing and promoting sustainable solutions like renewable energy and efficient electricity usage.

Smart energy transition, however, requires discussing different scenarios and solutions. Liquefied petroleum gas (LPG) has some different stories in Japan. More than 20,000 companies are operating LPG supply to final consumers in Japan. This also concerns city gas supplies. The current trajectory in Japan, aiming for carbon neutrality by 2050, focuses on reducing city gas carbon intensity through the gradual substitution of liquefied natural gas (LNG) with synthetic and bio-methane, alongside infrastructural and demand-side improvements. LNG is generally mixed with petroleum gas in order to constitute 45–46 megajoules. According to Akimoto and Kikkawa's (2023) "Methanation", hydrogen can be carefully injected into city gas directly. Hydrogen can be added to natural gas. Notably, Japan plans to achieve carbon neutrality without significant changes to its infrastructure, advocating for hydrogen-dedicated pipelines and lorries for final delivery. However, despite the cautious tone of the plan, critics highlight the absence of integrating hydrogen into existing city gas pipelines, a topic of significant debate within stakeholder circles. Also, nowadays in some parts of Japan such as Tokyo and Nagoya, some projects are now trying to use genuine hydrogen for direct combustions, injections, and

DOI: 10.4324/9781003471417-6

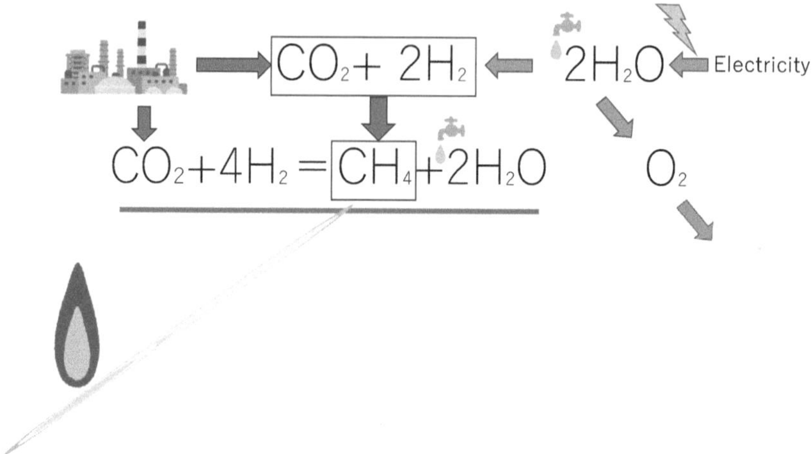

Figure 6.1 The basic concept of CO_2 utilisation for methanation.

Source: Author.

methanation. According to New Energy Foundation, Japan (2024a), the current city gas is good enough for environment, and one should remember to take their arguments seriously.

To ensure energy security and a stable supply of city gas, it is important to build a domestic production and supply system for synthetic methane and biomethane. In addition, stable long-term procurement of synthetic methane produced overseas is also important. Furthermore, after 2050, because of some policy shifts accordingly, there will be no prospect of long-term stable procurement of large volumes of synthetic methane and biomethane from domestic and overseas sources. It is because the Japanese government and some Japanese companies are trying to find places for creating e-methane. This occurred because of policy shifts by the Suga administration. Therefore, Japan plans to utilise LNG using carbon credits and carbon capture, usage, and storage (CCUS) for carbon recycling to ensure the stable supplies of city gas. This approach should, however, prioritise the continued stability of LNG procurement with a long-term vision extending beyond 2050.

6.2 Hydrogen Basic Plan and Green Transformation (GX) Net Zero by 2050

On 13 June 2020, the Gas Business System Review Working Group within the Ministry of Economy, Trade and Industry (METI) convened. During this meeting the Interim Review on Carbon Neutralisation of City Gas was compiled to the public. On the very next day, the Methanation Promotion Public–Private Council[1] was held, and a report was made on the interim review of this matter. Before this, in May 2020, discussions in the fuel sector towards carbon neutrality were held. This includes works on the interim report of the Public–Private Council to promote the introduction of synthetic fuel (e-fuel) and sustainable aviation. These reports served as a follow-up to the interim report of the Public–Private Council for promoting the introduction of SAF (sustainable aviation fuel), which will be used for aeroplanes. All the reports provide insights based on the current state of affairs explaining global trends in the related automobile and aviation industries under the 6th Energy Fundamental Plan. These discussions are closely related to addressing the implications of the Basic Hydrogen Strategy of Japan revised in June 2023. Additionally, they are a part of the preparation for the reorganisation of the Agency for Natural Resources and Energy (ANRE), which included the establishment of a new Hydrogen and Ammonia Division in the beginning of July 2023.

After nearly three years of deliberations, including the work of the Study Committee on the Future of the Gas Business towards 2050, which commenced in September 2020, subsequent reports were elaborated (ANRE, 2023a, 2023b, 2023c, 2023d, 2023e, 2023f, 2023g). This includes the draft version of a plan for the city gas carbon neutralisation issued in June 2023. The Carbon Neutral Action Plan by the Japan Business Federation[2] was issued in April 2024. Despite the extended duration, the progress in these debates has been relatively slow. This may stem from uncertainty surrounding the role of natural gas. Over the years, natural gas has been promoted as an effective measure for mitigating global warming due to its lower CO_2 emissions compared with oil or coal. However, there is a future scenario wherein the utilisation of natural gas is "evil", and natural gas would be treated like coal. Various perspectives exist, and as Japan continues to deliberate, prudence dictates a cautious approach in its actions. While Japan invokes the concept of carbon neutrality, it is essential to recognise the contemporary contexts that demand alignment with the ambitious targets set by Europe, the United States, or China. Rather than hastening to delineate the precise goal line, Japan must tread thoughtfully and judiciously.

In this context, synthetic methane, biomethane, and hydrogen are listed as temporary countermeasures, and CCUS and carbon credits are also addressed there in the context of synthetic methane that synthesises carbon dioxide and hydrogen. Hydrogen can be reliably secured, and Japan must respond wisely while monitoring the progress of the recently revised Basic Hydrogen Strategy nationally. Notably, the significance of biomethane has long been recognised, and the Energy Supply Structure Improvement Act[3] has already been mandated for the purchase of surplus biomethane. However, unfortunately, the amount purchased has decreased due in part to the impact of the support scheme (FIT/FIP).

The deliberation on Japan's Basic Hydrogen Strategy aligns with the actions outlined in the Green Transformation (GX), as established by the Basic Policy for GX – a Roadmap for the Next 10 Years. This policy was adopted in December 2020 under the leadership of the GX Executive Committee, chaired by then Prime Minister Suga. The 2020 Basic Policy provided a trajectory towards GX, which seeks to shift the existing industrial and social framework – predominantly reliant on fossil fuels – to cleaner energy sources. The goal is to achieve carbon neutrality in 2050 (see Sokołowski, 2022). This strategic approach, based on the imperative of ensuring a stable supply of energy, encompasses several key elements. These are promotion of energy conservation, renewable energy, hydrogen, and ammonia, as well as further use of nuclear power. By pursuing these avenues, Japan aims to contribute significantly to various sectors, including city gas infrastructure, transportation, and agriculture.

With respect to nuclear power, efforts will be made to develop and construct next-generation innovative reactors in order to ensure the sustainable utilisation of nuclear power in the future. While renewable energy serves as a crucial component of the Japanese energy mix, relying solely on it as the primary power source does not guarantee a stable energy supply. Therefore, the inclusion of nuclear power remains essential to meet Japan's energy demand effectively. In addition, to maximise the use of existing nuclear power plants, the current operational limit of up to 60 years will be maintained, with the possibility of extra extensions being considered. This aspect of the GX has garnered significant attention in the media as it stands as a pivotal element within the Energy Basic Policy.

6.3 Growth-Oriented Carbon Pricing Concept

The Energy Basic Policy contains another significant pillar of this basic policy, that is, the realisation and implementation of the Growth-Oriented Carbon Pricing Concept. It is a multifaceted framework that comprises several components, some of which have already materialised within the fiscal year 2020 (FY2020) budget bill. Specifically, this is referred to as the "technological support measures cost", as listed by the Ministry of the Environment and the METI (2024). As the Energy Basic Policy states, the necessary bills for the early realisation and implementation of this concept will be submitted to the next regular session of the Diet. Consequently, the relevant bills must be prepared prior to the parliamentary debate on the budget bill. It is highly likely that this will undergo deliberation in the Diet, resulting in introduction of GX Economic Transition Bonds.

Starting in FY2023, the Government of Japan has raised funds by issuing GX Economic Transition Bonds, aiming to upfront investment support with a goal of 20 trillion JPY[4] over the next 10 years. It is what is called Japan Climate Transition Bond (Cabinet Secretariat, 2023). According to the New Energy Foundation, Japan (2024b), this mechanism is based on two ideal pillars. First, the support must not only contribute to reducing CO_2 emissions but also to strengthen industrial competitiveness and promote economic growth. Second, the bonds should support projects that align with regulatory changes affecting corporate behaviours and demand-side practices using so-called "carbon pricing," which will affect those that lead to increased domestic human and material investments.

The redemption of GX Economic Transition Bonds or Japan Climate Transition Bonds will be facilitated through carbon pricing mechanisms. Carbon pricing improves the value of products and businesses that emit little or zero CO_2. However, the successful implementation of carbon pricing hinges on consideration of the development of alternative technologies and impact of CO_2 price on competitiveness. If not executed thoughtfully, it could burden companies and lead to the transfer of production overseas, having a negative impact on the Japanese economy. Immediate implementation of carbon pricing is ill-advised due to potential negative consequences. Instead, a phased approach is essential. Companies need time to proactively embrace the transition towards GX. By outlining a medium- to long-term trajectory towards this goal, companies can be incentivised to adopt necessary measures.

Based on these, Japan proposed two main mechanisms for carbon pricing: the Emissions Trading System (ETS), which focuses on companies that emit large amounts of CO_2, and the Carbon Levy, a broader incentive framework supporting GX. Regarding ETS, the GX League, comprising companies committed to proactive GX efforts, has seen participation from approximately 600 companies since November 2020. METI, which serves as the GX League's secretariat, plans to proceed with the study and trial runs, with the full-scale operation from FY2026. Separately, for power generation companies (where it is difficult to transfer production overseas and there are alternatives like renewable energy and nuclear power), the renewable energy levy (FIT) has peaked. Drawing lessons from the European Union, emission allowances will be allocated through paid auctions, commencing in 2033, coinciding with the anticipated release of emission allowances.

Earlier, in 2028, a carbon emission charge will be introduced. This will occur after a five-year preparation period. The plan entails initially introducing the tax at a low cost for fossil fuel importers and raising it. While the support budget of approximately 20 trillion JPY may be captivating, it is crucial to recognize that this support is integrated with regulations and institutional measures. Therefore, close attention to future trends is necessary to ensure effective implementation.

With respect to synthetic methane within Japan's GX policy, in order to utilise it on a large scale, it is essential to establish further mass production by 2030 or 2050. To realise this goal and propel Japan's GX initiatives, several policy measures are needed. First, further technological development is required. Second, subsidies should be offered to companies and research institutes committed to advancing methanation. To achieve large-scale methanation, the formation of a consortium comprising companies emitting substantial amounts of hydrogen and carbon dioxide is anticipated. Additionally, establishing a new corporate alliance would enhance our capabilities, particularly in relation to cold energy and alternative fluorocarbons. Therefore, the GX League should invite smaller partners to the consortium.

Currently, the GX League stands as a substantial institution, yet the technologies it employs are inherently compatible, so having smaller business alliances take the initiative in creating one could be an important policy. In particular, when a consortium is formed using the GI Fund, GX Economic Transition Bonds or Japan Climate Transition Bonds, etc., confidentiality obligations arise, making it difficult to collaborate with people outside the consortium. As a next step, the introduction of a new fund – for instance, one that can induce, for example, the reorganisation of a consortium using a new propanation framework, there will be a drive towards propanation research and development (R&D), which will eventually lead to large-scale development of synthetic methane. Some consortiums will be successful to promote methanation.

Moreover, the new breakthroughs remain possible with fresh implementations. It is important to consistently allocate financial resources by creating a dedicated fund. While the GI fund has reinvigorated methanation R&D by major city gas companies, Japan's industries still express dissatisfaction and seek additional support. It started in April 2024 with issuances of bonds by the Japanese government. The GI fund is not a large fund. Therefore, Japan Climate Transition Bond and other funds can support the total schemes. Japan started the issuance of 20 trillion JPY of GX Economy Transition Bonds.

Regarding policy measures to achieve mass production of city gas, the development of carbon-neutral gas technology is slated to align with the Nationally Determined Contributions (NDCs) targets for 2030 and the ultimate goal of carbon neutrality by 2050. Multiple policy responses can contribute to achieving this objective. Public awareness plays a crucial role in achieving NDC targets. Some individuals may struggle to envision the artificial synthesis of methane, therefore, conducting awareness-raising activities is needed. Additionally, emphasising gas conservation is essential.

6.4 Synthetic Methane (E-methane) Production

The methane production using the new technology has come to be called "synthetic methane" or "e-methane" in Japan. While the term "e-methane" is not yet widely adopted, the concept of synthetic methane produced from renewable or clean resources is gaining traction globally. However, it is important to differentiate between the specific term e-methane and the broader concept of synthetic methane.

With respect to the pricing method for long-term synthetic methane supply contracts, while the specific pricing approach has not been determined, it is essential to consider the transition period leading up to 2050. It should also be associated with specific Japanese safety regulations and standards in the future. Synthetic methane supply prices are assumed to be higher than LNG import prices under long-term contracts. Consumers are advocating for the adoption of synthetic methane, which shares similarities with LNG but boasts a "carbon-neutral value".

Regarding the price difference with LNG during the transition period, it is often considered as a policy response within the timeline of achieving NDC in 2030 and achieving carbon neutrality in 2050. We should always pay attention to the price difference between e-methane and LNG from now on. It means building a mechanism to compensate for the price difference based on the price difference between e-methane and LNG. The industry's cost target is projected to be 120 JPY/m^3 by 2030 and 50 JPY/m^3 by 2050. It is widely anticipated that by these dates, the price of e-methane will align with the historical LNG price. However, this scenario hinges on technological advancements in e-methane production, supported during the transition period to offset price difference compared with LNG price. Moreover, investing in e-methane from an early stage, driving decarbonisation, and strategically reducing manufacturing costs can lead to increased efficiency of methanation equipment and a reduction in the cost of renewable energy electricity. The aim of reducing manufacturing costs through securing suitable manufacturing sites is a critical objective. Considering the target synthetic methane price in conjunction with trends in LNG prices and city gas sales, and acknowledging the 1% e-methane introduction target for 2030, it is necessary to establish a regulatory framework by around 2025. This deadline is crucial for making decisions on specific projects. Anticipating a future where synthetic methane prices align with LNG price trends and city gas sales, Japan aims to capture overseas markets. By entering the e-methane market at an early stage, Japan has the potential to become a leader in the global e-methane industry. To further advance the adoption of e-methane, concerted efforts should focus on strengthening both domestic and international distribution networks. By strategically addressing these aspects, countries can position themselves for a sustainable and resilient synthetic methane supply chain.

Currently, in Japan, small- and medium-sized businesses, not to mention major companies, face challenges in independently producing e-methane. As a result, they must procure e-methane from external sources. To address this issue, it would be beneficial to prioritise the establishment of reloading facilities to stimulate distribution. If reloading capabilities are implemented across different regions in Japan, lorries can be used, which will contribute to the nationwide expansion of the supply and increase business opportunities for distribution for businesses.

In the context of raw materials from overseas such as congested methane with differentiated energy content, ensuring stable procurement and securing city gas supply was, and still is, important. Beyond the conventional practices associated with LNG procurement, it is crucial to formulate policy responses that address the acquisition and safeguarding of methane from international sources. This perspective is particularly significant when viewed from the perspective of energy security. By implementing methanation overseas and using Japanese technology, gas-producing countries can continue to export e-methane. Japan should transition away from the structure of the fossil fuel importing framework and cooperate to successfully liquefy and export e-methane globally.

6.5 Policy Framework for E-methane Production

The 6th Basic Energy Plan of Japan (2021) marks the 10th anniversary of the Great East Japan Earthquake and the TEPCO Fukushima Daiichi Nuclear Power Plant accident. The steady progress of Fukushima's reconstruction and the priority given to safety as the foremost consideration serve as fundamental premises for advancing energy policy. In addition, under the 2021 energy policy, Japan will prioritise energy security, with safety as the premise, and energy supply at low cost by improving economic efficiency. This is following the "S+3E" perspective (safety plus energy security, economic efficiency, and environment), which aims to achieve these pillars along with protecting the environment. Moreover, within the framework of the 6th Energy Basic Plan, Japan charts a strategic course for energy policy towards realising the 2050 carbon neutrality, as it introduces the new greenhouse gas emission reduction goals. Furthermore, while promoting climate change mitigation measures, Japan emphasises the importance of ensuring stable energy supply and reducing energy costs. These initiatives are essential for overcoming the challenges faced by Japan's energy supply and demand structure.

The new outlook for energy supply and demand calls for a 46% reduction among top three city gas companies, Tokyo Gas, Osaka Gas, and Toho Gas, in greenhouse gas emissions in FY2030 compared with FY2013. In light of the Japanese government announcement that it will continue to strive for a higher goal of 50%, though it officially aims to reduce emissions by only 46%. The ambition is to overcome various challenges on both supply and demand sides in promoting thorough energy conservation and expansion of non-fossil energy. This shows what kind of energy supply and demand will be projected based on the current scenario. When implementing measures to achieve this ambitious outlook, Japan will strengthen measures to ensure that stable supply is not disrupted. The degree and timing of implementation must be carefully considered. For example, at a stage before non-fossil power sources are fully introduced, if measures are taken to curb fossil power sources in the future, the stable supply of electricity could be disrupted. Furthermore, as carbon-neutral policies continue to change dynamically, I believe that policy support is needed to promote upstream development that makes it easier to conclude long-term situations.

6.5.1 Natural Gas as a Fossil Fuel with Low CO_2 Emissions

Japan will promote decarbonisation efforts using renewable energy by popularising natural gas. This means that J-Credits will be used with the understanding of society. J-Credits will work for the deduction of emission of CO_2. Ultimately, Japan aims to achieve net-zero carbon emissions by 2050. Uncertainty in LNG demand is also a concern. Although it is difficult for Japanese companies to terminate long-term LNG contracts, term contracts are decreasing. This does not mean that term contracts will become less important. Specific examples of advanced natural gas utilisation are fuel conversion and smart energy networks. On the demand side, we will thoroughly reduce energy and CO_2 emissions through advanced use of natural gas. Securing carbon-neutral LNG is still in the process of making rules, and it is not yet at the stage where firm offsets can be implemented. Rather than just promoting the energy savings of natural gas and its resulting reduction in CO_2, it is possible to further promote environmental value by combining natural gas with voluntary credits.

6.5.2 Next-Generation Heat Demand

Comparing hydrogen, ammonia, and e-methane as means of decarbonising heat demand, e-methane's feature is that it can utilise existing infrastructure. Social costs can be minimised by utilising existing consumer equipment. The value provided by e-methane is decarbonisation in the heat field. This will contribute to achieving carbon neutrality in heat demand, which accounts for 60% of Japan's energy demand, thereby reducing additional social costs. This means that existing infrastructure can be used effectively.

6.5.3 Ensuring Energy Security

Like Southeast Asia, Japan will seamlessly achieve carbon neutrality by converting coal to natural gas and then converting it to e-methane. For Japan, this is extremely important to achieve a 1% e-methane introduction by 2030. In Japan, Tokyo Gas Co., Ltd., the largest city gas company in Japan, is conducting demonstration tests at a facility in Tsurumi Ward, Yokohama City. Although it is small scaled at 12.5 Nm^3/h, it is promoting technological development that includes the introduction of innovative technology, and the company expresses its willingness to eventually demonstrate a medium-scale test at around 100 Nm^3/h. Along with Tokyo Gas Co., Ltd., Osaka Gas Co. Ltd., the second largest city gas company in Japan, performs similar demonstration tests in Konohana Ward, Osaka City. Its main feature is SOEC for the methanation, which is close to the plan of Tokyo Gas Co., Ltd.

6.5.3.1 Price-Difference Policy

As a policy response, Japanese people should continue to pay close attention to the price difference between e-methane and LNG within the time frame of achieving the goal of realising the LNG. First, Japan will build a system to compensate for the price difference based on the price difference between e-methane and LNG. The industry's cost target is 50 JPY/Nm^3 in 2050, and many believe that by then the price will be at the same level as the previous LNG price, but LNG prices will skyrocket. Japan should keep this price level of a Nm^3 that stands for a competitive value of carbon neutral energy.

However, the situation changes when a technological breakthrough occurs in the production of e-methane. Currently, the e-methane production cost in 2030 is aimed at approximately 120 JPY/Nm^3, and from the introduction of e-methane to its widespread use, a support system is needed to compensate for the difference with the LNG price. I think that Japan will encourage investment in e-methane from an early stage to drive decarbonisation. In addition, Japan should aim to reduce manufacturing costs by leveraging larger and more efficient methanation equipment, minimising renewable energy electricity costs and securing suitable manufacturing sites.

6.5.3.2 Expansion Drill

Considering the 1% e-methane introduction target for 2030, which is the target for synthetic methane price based on trends in LNG prices and city gas sales, rules need to be established by around 2025, which is considered the deadline for making decisions on specific projects. Therefore, it is important for countries to cooperate in negotiations between countries. Surprisingly, it seems that the day will come soon when synthetic methane prices will be comparable to trends in LNG prices and city gas sales. Furthermore, if Japan can capture overseas markets and become able to enter the e-methane market from an immature stage, Japan may be able to lead the overseas e-methane market.

From the perspective of stable supply and energy security, the expansion of domestic methanation and the diversification of supplier countries and the entry of foreign companies in overseas methanation are expected. Various efforts are being made overseas to produce and supply hydrogen. Policy measures are also needed to build a robust synthetic methane supply system. With pure synthetic methane, there is significantly less need to worry about the possibility of rollover within the terminal depending on the production area and composition, unlike with conventional LNG, so we should actively consider this as an advantage of e-methane.

Japan should also aim to strengthen domestic and international distribution. In Japan, it is difficult for small- and medium-sized businesses, not to mention major companies, to manufacture e-methane on their own, so they must procure e-methane from somewhere else. Therefore, it would be beneficial to give preferential treatment to the construction of bases with reloading functions to stimulate distribution. If reloading becomes possible in various parts of Japan, conventional lorries can be used, which will contribute to the nationwide expansion of the supply of synthetic methane, and e-methane will increase business opportunities because of new businesses in distribution. This will give momentum to the new distribution.

6.5.4 CO_2 Emissions of E-methane

It is important to set the CO_2 emissions count of e-methane to zero under the domestic rules of what they call SHK (*Santei*, *Hōkoku*, *Kōhyō*, which translates to "calculation, reporting, and making them public"). An agreement on CO_2 emissions counting will be reached between governments based on a private-sector level agreement. For carbon offsets to ensure reliable and transparent credit investment predictability, it is necessary to clarify the requirements for voluntary credits, such as GX-ETS, that are positioned in the system.

Japan will achieve seamless carbon neutrality by continuously carrying out decarbonisation centred on the transition from natural gas to e-methane and biomethane. The value of e-methane and biomethane, which are central in the thermal energy field, will provide the basic infrastructure of city gas. It will be contributing to the decarbonisation of heat demand while suppressing social costs. In this light, a Japanese energy company Osaka Gas, to contribute to carbon neutrality in Asia, will focus on technology development, practical application, and supply chain construction, and aim to introduce 60 million m^3 of e-methane at 1% in 2030. Towards decarbonisation of gaseous energy, e-methane will be the main fuel due to the possibility of securing needed volume. That is why Osaka Gas aims to secure stable and inexpensive e-methane by building supply chains both domestically and internationally, including large-scale procurement through overseas projects, which it has already begun considering.

E-methane is one of the hydrogen carriers, and if domestic methanation is promoted, the use of hydrogen as a raw material for e-methane will progress, contributing to the expansion of domestic hydrogen demand. In terms of technology, through Sabatier methanation, Japan will achieve the introduction of 1% e-methane in 2030 and aim to expand the spread of e-methane through further scale-up. Furthermore, e-methane is necessary for carbon neutrality in 2050. According to Chen, Chen, and Xia (2014), on theirthesis "Direct Synthesis of Methane from CO_2-H_2O Co-Electrolysis in Tubular Solid Oxide Electrolysis Cells,"for the development of solid oxide electrolysis cell (SOEC) methanation will be able to synthesise methane from water and CO_2 directly and efficiently. Japan will take on the challenge of developing basic technology with the help of the Green Innovation Fund, aiming for commercialisation around 2040. In Japan, biomass has the problem of limited availability, but by utilising the surplus CO_2 in biogas through biomethanation, we can maximise its potential and contribute to the expansion of domestic energy production and local consumption.

6.5.5 Cost Down of Methanations

As of 2030, Japan aims for 120 JPY/Nm^3 through Sabatier methanation. This would cost approximately 150,000 JPY per ton when converted to the current price of natural gas. After that, we aim to achieve a reduction of 50 JPY/m^3 in 2050 through the practical application and scale-up of highly efficient SOEC methanation. According to Zheng, Y. et al. (2017), there are some possibilities that much higher temperature than that of incumbent e-methane production can produce more e-methane as long as they rely on SOEC. Similarly, if we replace this with the price of natural gas, it will be 60,000 JPY per ton, which is almost the same level as the natural gas price before the situation in Ukraine worsened. Securing cheaper renewable energy is also an important issue. In order to manufacture e-methane at a low cost in major overseas e-methane projects, it is important to secure renewable electricity, water, and CO_2 cheaply and stably near the existing infrastructure of LNG liquefaction terminals. In collaboration with partner companies, we are investigating the possibility of e-methane manufacturing business in North America, South America, Australia, and Southeast Asia that meet these conditions, and Japan plans to make an investment decision on the first project in 2025.

6.5.6 Efforts to Maximise the Use of Biogas

Currently in Japan, the surplus CO_2 in biogas accounts for about 40% of the total, but by methanating this carbon dioxide and hydrogen, the methane concentration can be raised from 60% to 95%, 1.5 times higher than before. Japanese energy companies are proceeding with efforts to produce methane. This technology will maximise biogas, which is a locally produced and locally consumed energy. They learn from Europe. The Japanese government believes that this will contribute to the decarbonisation of local gas utilities, which have difficulty procuring e-methane on a large scale from overseas.

6.6 Japanese Companies and E-methane

There are several Japanese companies engaged in different activities related to e-methane. For instance, Osaka Gas. is currently considering new methanation methods using unused biomass using CCUS in Malaysia. It methanates hydrogen and carbon mono-oxide generated during the process of high-temperature gasification of biomass. Conventional methanation requires a large amount of electricity during electrolysis, but this method uses almost no electricity. This makes it possible to produce e-methane without being affected by the price of renewable electricity.

Mitsubishi Corporation is considering multiple overseas projects, including a four-party joint venture with Tokyo Gas, Osaka Gas, and Toho Gas, but the earliest project could be started within one or two year(s). Toho Gas, the third-largest city gas company in Japan, is developing its business as an energy business with roots in the region, mainly in Aichi, Gifu, and Mie prefectures in Japan. This area, where Toyota's headquarters is based, is one of Japan's leading industrial clusters, focusing on the automobile industry, and is characterised by the high demand for gas from industries that are active on a global scale. These customers have very strong needs for gas decarbonisation.

In 2021, Toho Gas announced its commitment to become carbon neutral in 2050, demonstrating to a wide range of stakeholders the group's determination and future direction. The scenario for achieving carbon neutrality requires not only the accumulation of technological innovations but also the cost and time needed to overcome it. It is extremely important to control CO_2 emissions from the ground up and reduce cumulative CO_2 emissions by switching fuels from coal and oil to city gas, utilising eco-generation, and thoroughly using advanced energy. It is anticipated that that smooth transition to carbon neutrality will be conducted in a reasonable manner while reducing social burden and maintaining customer convenience. It is also worth mentioning that Idemitsu Oil is trying to produce ammonium in the city of Yokkaichi, Mie prefecture, among other places.

An effective scheme would be to manufacture e-methane overseas, where the cost of procuring renewable energy used for hydrogen production is relatively low, and import it into Japan using existing infrastructure such as LNG shipping terminals and transport ships. It aims to make an investment decision in FY2025, with a view to starting the introduction of 1% or more e-methane in 2030, including conducting business feasibility studies with Toyota Tsusho and Total Energy. To date, it has been using surplus biogas generated at a sewage treatment plant in the city of Chita, Aichi prefecture, in our efforts to produce methanation in Japan. Approval is scheduled for the second half of 2023.

It is expected to be the first time in Japan to use e-methane produced by Toho Gas as a raw material for city gas, and together with local industrial users like Aisin or Denso, it will convert CO_2 emitted from inland factories into city gas in production plants. It will be a regional circulation model where it will be transported overland and methanated. There are cost issues related to CO_2 liquefaction and transportation. This is an initiative related to the separation and recovery of CO_2. Carbon dioxide generated from a large number of emission sources, such as industrial customers, is efficiently captured and used for fixed use in methanation, materials, etc., which can be used as fuel. I believe that CCUS will become a strong option for Toho Gas in the future.

In order to efficiently recover CO_2 from industrial customers, Toho Gas has installed demonstration equipment for different methods of separation and recovery within its laboratory. It is currently building and evaluating its performance. An initiative related to CO_2 separation and recovery near LNG terminals is Cryo-Capture, which recovers CO_2 from exhaust gas from large-scale factories in the bay area. This is Cryo-DAC, which aims to capture CO_2 from the atmosphere. It is also actively working to develop efficient CO_2 separation and recovery technology that utilises the cold energy that is exhausted after the gas is transported by sea. These technological developments are proceeding with support from the government, and in the pilot demonstration phase under the Green Innovation Fund, it is planning to demonstrate the production of e-methane using CO_2 collected at LNG terminals.

6.7 Discussion

There are no rules in place to distinguish between city gas emissions using liquefied natural gas and e-methane. To ensure that users are rewarded even if they choose the more expensive e-methane, we are implementing measures such as greenhouse gas protocols, including the use of certificates proving the environmental value of e-methane, as well as establishing rules for the domestic SHK system and GX League. There is a need for a system that allows for the recognition of different emission factors for city gas in international climate change initiatives, etc., and support from the government in tandem with the private sector is needed to realise this.

To meet the needs of customers, whose expectations for e-methane are increasing, the Japanese government believes that policy support that helps companies maintain their competitiveness is essential. From the perspective of establishing rules, it will be important to develop a system that allows e-methane users to be evaluated as having zero emissions. In addition, public and private sectors need to work together to develop CO_2 counting rules between manufacturing and user countries, which will be the basis for international rule development and policy support.

From the perspective of policy support, it is important to reduce the cost burden on users and increase the predictability of investment in manufacturing and import projects. It is necessary to develop cost support measures to compensate for the difference. In particular, generous support is needed for first movers who face high uncertainties about the future and the risks involved in making large-scale investments. There is a potential for fuel conversion from coal and oil by 19 million tons in terms of reduction. Efforts should be made to reduce CO_2 emissions across the board by switching fuels, saving energy, and reducing CO_2 emissions in individual devices, and using smart energy to link these efforts.

One of the major contributions at recent energy forums, largely involving academics, centred on the need for the development and coordination of framework on the use of synthetic methane. This stems from the lack of clear regulations regarding CO_2 emissions associated with synthetic methane at both the national and corporate levels. This issue extends beyond synthetic methane, encompassing synthetic fuels as well as carbon recycling. For example, consider someone who captures CO_2 emissions from a fossil fuel boiler by reducing its release into the atmosphere, then selling that CO_2 as a raw material for synthetic methane production. Proponents of synthetic methane naturally claim that it is "CO_2-free", while those capturing CO_2 are doing this to reduce their own CO_2 emissions. If both claims are accepted, it results in double counting, highlighting the complexity of establishing effective regulations in this regard.

Finally, from the perspective of stably procuring and securing city gas raw materials from overseas, in addition to the conventional procurement and securing of LNG, it is also important to consider policy responses to the procurement and securing of e-methane from overseas. This is also important from the perspective of energy security. Increasing the self-sufficiency rate is important to some extent, but by implementing methanation overseas and using Japanese technology, gas-producing countries can continue to export e-methane, allowing them to maintain friendly relations between countries as before. Countries will break away from the structure of exporting fossil fuels and work together to liquefy and export e-methane. The injection of synthetic methane and biomethane has a lower calorific value than the general way of using LNG. It goes into city gas pipelines. It will be a way of reducing the standard calorific value. It is a more appropriate calorie system than the band system, which has many allowances. It will lower the amount of heat, conduct preliminary verification, and then determine the optimal amount of heat.

6.8 Conclusion

Methanation, a promising technology for making clean city gas (e-methane), is crucial for achieving a zero-carbon society. Mass production by 2030 and 2050 is key for widespread adoption of this technology in Japan. There are various possible policy responses to achieve this and promote Japan's GX. Among them are subsidies to companies and research institutes that want to focus on methanation, and to support R&D that will contribute to mass implementation. To implement large-scale methanation, Japan first needs to form a group of companies that emit large amounts of hydrogen and carbon dioxide. It is ideal to form a new corporate alliance to expand connections with the use of cold energy and alternative fluorocarbons. Furthermore, achieving GX necessitates fostering public awareness. With this support for GX, significant progress can be made towards fulfilling NDC, ultimately paving the way for Japan to reach carbon neutrality by 2050.

Notes

1 In Japanese: メタネーション推進官民協議会 [metane-syon suisinn kanmin kyougikai].
2 In Japanese: 日本経済団体連合会 [nippon keizai-dantai rengōkai].
3 In Japanese: エネルギー供給事業者によるエネルギー源の環境適合利用及び化石エネルギー源料の有効な利用の促進に関する法律 [enerugii kyoukyuu jigyousha niyoru enerugii gen no kankyou tekigou riyou oyobi kaseki enerugii genryou no yuukou na riyou no sokushin ni kansuru houritsu].
4 In late June 2024, 1 USD equalled approximately 160 JPY.

References

Akimoto, K. and Kikkawa, T. (2023) "Methanation", *Energy Forum.*

ANRE, Comprehensive Natural Resources and Energy Investigation Committee, Electricity and Gas Business Subcommittee, Electricity and Gas Basic Policy Subcommittee (2023a), "26th Gas Business System Review Working Group", 8 February 2023.

ANRE, Comprehensive Natural Resources and Energy Investigation Committee, Electricity and Gas Business Subcommittee, Electricity and Gas Basic Policy Subcommittee (2023b), "27th Gas Business System Review Working Group", 13 March 2023.

ANRE, Comprehensive Natural Resources and Energy Investigation Committee, Electricity and Gas Business Subcommittee, Electricity and Gas Basic Policy Subcommittee (2023c), "28th Gas Business System Review Working Group", 18 April 2023.

ANRE, Comprehensive Natural Resources and Energy Investigation Committee, Electricity and Gas Business Subcommittee, Electricity and Gas Basic Policy Subcommittee (2023d), "29th Gas Business System Review Working Group", 16 May 2023.

ANRE, Comprehensive Natural Resources and Energy Investigation Committee, Electricity and Gas Business Subcommittee, Electricity and Gas Basic Policy Subcommittee, 30th Gas Business System Review Working Group (2023e), 23 May 2023.

ANRE, Comprehensive Natural Resources and Energy Investigation Committee, Electricity and Gas Business Subcommittee, Electricity and Gas Basic Policy Subcommittee (2023f), "31st Gas Business System Review Working Group", 13 June 2023.

ANRE, Comprehensive Natural Resources and Energy Investigation Committee, Electricity and Gas Business Subcommittee, Electricity and Gas Basic Policy Subcommittee (2023g), "32nd Gas Business System Review Working Group", 9 November 2023.

Cabinet Secretariat (2023) "Japan Climate Transition Bond Framework". Available at: www.cas.go.jp/jp/seisaku/gx_jikkou_kaigi/ikousai/pdf/climate_transition_bond_framework_eng.pdf (Accessed: 13 June 2024).

Chen, L., Chen, F. and Xia, C. (2014) "Direct Synthesis of Methane from CO_2-H_2O Co-Electrolysis in Tubular Solid Oxide Electrolysis Cells", *Energy & Environmental Science*, 7, pp. 4018–4022.

New Energy Foundation, Japan (2024a) "Energy Conservation in the Transport Sector". Available at: www.nef.or.jp/keyword/ta/articles_to_03.html (Accessed: 13 June 2024).

New Energy Foundation, Japan (2024b) "Energy Efficiency Technologies in the Commercial Sector". Available at: www.nef.or.jp/keyword/abc/articles_g_02.html (Accessed: 13 June 2024).

Ministry of Economy, Trade and Industry (METI) (2024) "Transition Finance". Available at: www.meti.go.jp/english/policy/energy_environment/transition_finance/index.html (Accessed: 13 June 2024).

Sokołowski, M.M. (2022) *Energy Transition of the Electricity Sectors in the European Union and Japan: Regulatory Models and Legislative Solutions*. Cham: Palgrave Macmillan.

Zheng, Y. *et al.* (2017) "A Review of High Temperature Co-Electrolysis of H_2O and CO_2 to Produce Sustainable Fuels Using Solid Oxide Electrolysis Cells (SOECs): Advanced Materials and Technology", *Chemical Society Reviews*, 46, pp. 1427–1442.

7 Could the Mobility as a Service (MaaS) Contribute to the Achievement of Japan's Net-Zero Goal in the Transport Sector?

Jun Yamashita

7.1 Introduction

Mobility as a service (MaaS) apps have been developed and used worldwide since their commercialisation in the early 2010s. According to Mladenović (2021), MaaS is a service that provides users with a "one-stop shop" or "mobility platform" via digital interfaces to access information, book, and pay for a variety of mobility services. Notably, UbiGo, piloted in Gothenburg in 2013, is widely recognised as the world's first MaaS app. The use of MaaS apps has primarily expanded in developed countries, with notable examples including "Whim" in Finland and "Moovit" in Israel. While Whim has also been introduced in Japan, domestically produced MaaS apps, such as "My Route", have been used since the late 2010s. It is worth noting that while MaaS was first introduced in Europe, its use on a global scale is still limited (Santos and Nikolaev, 2021).

MaaS apps were specifically developed to promote transit-oriented development in Europe, with a strong focus on transport policy. It is important to note that the position of MaaS in transport policy is largely due to the Intelligent Transport Systems (ITS) EU Directive enacted in 2010. This Directive (2010) aimed to establish a framework that supports the coordinated and coherent deployment and use of cross-border ITS within the European Union. The Directive sets out the general conditions required to achieve the stated goal (Article 1(1)). The use of ICTs and Internet of Things (IoT) in transport systems has been enhanced as a result of this enactment. An example of such use is the MaaS.

The European Union's 2020 Sustainable and Smart Mobility Strategy suggests the use of ITS as a solution to environmental issues, such as air pollution resulting from traffic congestion and GHG emissions. In the same line with the strategy, the ITS Directive has been amended to include environmental sustainability. The amendment aimed to address a range of emerging technologies, including those related to connected cars, automated driving, on-demand mobile apps for traffic information, and reservation and ticketing services that combine different modes of transport, such as car sharing and public transport.

With regards to smart transport policy, Sochor *et al.* (2018) classified MaaS into different stages. The first level involves individual transport services providing independent information without integration (Table 7.1). At the second level, a

DOI: 10.4324/9781003471417-7

Table 7.1 Typology of MaaS

MaaS levels	*Integration level*	*Functions*
Level 4	Integration of societal goals	Policies, incentives, etc.
Level 3	Integration of the service offer	Bundling/subscription, contract, etc.
Level 2	Integration of booking and payment	Single trip – find, book, and pay
Level 1	Integration of information	Multimodal travel planner, price information
Level 0	No integration	Single, separate services

Source: Sochor *et al.* (2018).

MaaS app integrates information on prices, travel times, and distances of several transport services, making it searchable. At the third level, an app integrates the reservation, payment, and ticketing of multiple transport services. According to this typology, at third level, various transport services may be provided through a subscription or bundling package. At the fourth level, the integration of urban and transport planning policies can be accomplished through collaboration between the national government, local authorities, and operators. It is suggested that the most advanced MaaS could be connected to transport policy.

Extensive research has been conducted on MaaS in conjunction with the proliferation of MaaS apps and the advancement of MaaS-based transport policies. MaaS is being approached from various perspectives, but it can be contextualised within the framework of sustainability transitions. This is because MaaS aims to establish sustainable and intelligent transport systems by utilising information and communication technology (ICT)-centric solutions (Smith, Sochor, and Sarasini, 2018; Köhler *et al.*, 2019). Liu *et al.* (2017) identified three main technology areas that support urban infrastructure, including transport systems: the IoT, data mining technologies, and mobile wireless networks such as vehicle positioning systems. Lauri, Shimpo, and Sokołowski (2023) discussed contact tracking applications and how they may aid in the development of smart cities, benefiting public transportation while also providing valuable insights for city management. The MaaS concept is related to all of these technologies. Banister and Stead (2004) note that the impacts of ICT solutions on environmental sustainability are complex. Chawla, Shimpo, and Sokołowski (2022) advocated for developing green artificial intelligence (AI) rather than simply "developing AI" with the aim of utilising it for climate policy, energy transformation, and achieving climate neutrality. Regarding the environmental sustainability, MaaS solutions have also been discussed by several researchers (Arias-Molinares and García-Palomares, 2020; Kriswardhana and Esztergár-Kiss, 2023; Maas, 2022; Pangbourne *et al.*, 2020; Rey-Moreno, Periáñez-Cristóbal, and Calvo-Mora, 2023; Santos and Nikolaev, 2021).

Notably, in order to establish MaaS as a new technology in society, users' attitudes and behaviour need to change from use of private cars to that of other sustainable transport modes, such as public transport. According to Kamargianni *et al.* (2018), only 13% of permanent car users continue to use their private car after the introduction of MaaS, while 35% intend to switch to public transport. Moreover, 17% of the participants stated that they had switched to cycling or walking. These findings indicate that it is not possible at this stage to make any conclusive statements about changes in attitudes towards more environmentally friendly modes of transport following the implementation of MaaS.

It also can be challenging to make definitive conclusions about the behavioural changes resulting from the attitude shift caused by MaaS. While Hensher *et al.* (2020) demonstrated that MaaS may encourage users to reduce their car usage, Hesselgren, Sjöman, and Pernestål (2020) argued that the impact of MaaS on user behaviour is limited. It could be argued that there is currently insufficient research to make definitive claims about the impact of MaaS on attitudes and behaviour. Therefore, it may be beneficial to conduct further studies in this area.

Although the introduction of MaaS has shown some promise in reducing car use, further research may be needed to confirm its impact on increasing public transport use. As noted by Maas (2022) in his review of the impact of MaaS on the transport system, there are claims that MaaS could potentially address major societal challenges related to transport, such as improved accessibility, reduced congestion, and environmental sustainability. The implementation of MaaS in Sydney has been observed to lead to a decrease in private car usage. These findings suggest that reducing cars has the potential to alleviate congestion. As a result, there has been an increase in the use of public transport and car sharing. The studies conducted by Hensher, Ho, and Reck (2021) and Kamargianni *et al.* (2016) provide further support for these findings. The use of MaaS has also shown potential for increasing walking, especially for short distances (Lyons, 2020).

The MaaS might have both positive and negative implications for transport. According to Kriukelyte (2018), the implementation of MaaS could potentially lead to a decrease in car usage and an increase in the use of public transport. However, some individuals who previously used public transport may opt for alternative modes of transportation, such as taxis, car sharing, cycling, or walking. This means that the introduction of MaaS does not always result in an overall increase in public transport usage. As a result, it has been suggested that some individuals may choose to travel longer distances by private car after the introduction of MaaS (Hörcher and Graham, 2020). Additionally, there is a possibility that the use of public transport may increase, which could lead to overcrowding on buses and trains, and, in turn, a reduction in use of public transport due to overcrowding.

In addition, further research is also needed to assess the environmental sustainability of implementing MaaS, particularly with regards to its potential to reduce GHG emissions. Studies on MaaS often assume that it can alleviate congestion in areas where traffic is heavy by encouraging a modal shift from private cars to public transport, walking, or cycling. It is widely acknowledged that reducing GHG emissions in the transport sector can have a positive impact on environmental sustainability. Simulation analyses have suggested that the introduction of MaaS may have the potential to reduce GHG emissions through car-sharing and bike-sharing (Becker *et al.*, 2020). However, the impact on private car use and congestion reduction remains uncertain, and the sustainability of MaaS is still a matter of debate (Pangbourne *et al.*, 2020). Furthermore, while some case studies have been conducted, there is limited evidence to establish a clear relationship between MaaS and environmental sustainability (D'Amico, 2023). The impact of MaaS on GHG emissions in a specific region is a topic that is still being discussed. In fact, as noted by Wittstock and Teuteberg (2019), MaaS can have both positive and negative effects on GHG emissions.

In the light of the discussed studies, this chapter aims to identify changes in user attitudes resulting from the implementation of MaaS and to assess its influence on environmental sustainability. These purposes are examined using empirical evidence from MaaS projects implemented by the Japanese government. Section 7.2 outlines the smart city initiatives and highlights their defining characteristics. In Section 7.3, the impact of MaaS on users' attitudes and environmental sustainability is clarified, drawing on project reports from smart city initiatives. Section 7.4 delves into the reasons behind users' attitudes and the effects of MaaS. Finally, in Section 7.5, policy implications for future smart city initiatives towards sustainability transitions with MaaS in Japan are explored.

7.2 Smart City Initiatives in Japan

In 2016, the Japanese government presented the concept of "Society 5.0" as part of the 5th Science and Technology Basic Plan. Society 5.0 is the super-smart society (Shimpo, 2018) that creates a human-centred society that balances economic advancement with the resolution of social problems by highly integrating cyberspace and physical space through the use of ICTs. Society 5.0 is a concept that shares similarities with the "Industry 4.0" initiative proposed by the German government and the Advanced Manufacturing Partnership proposed by the US government. It represents the next stage of societal evolution following hunting society (Society 1.0), agricultural society (Society 2.0), industrial society (Society 3.0), and information society (Society 4.0).

7.2.1 Overview

In 2019, the Japanese government launched various smart city initiatives based on the integrated innovation strategy. In this strategy, smart cities are regarded as the ideal location to realise the concept of Society 5.0. This strategy presents a clear vision for the future of smart cities, which involves sharing basic concepts and best practices of data collaboration infrastructures among smart cities across different countries. The ultimate goal is to solve problems faced by cities and regions worldwide, while fostering global collaboration and cooperation. To strategically promote convenient and comfortable city planning for the future, utilising new technologies such as IoT, the first step is to build a government-wide smart city infrastructure in cooperation with various ministries and agencies.

To implement various smart city projects in different fields, collaboration between the public and private sectors is proposed in the 2019 strategy. A public–private partnership platform should be established to promote this horizontal collaboration with confidence. The Cabinet Office (CO), the Ministry of Internal Affairs and Communications (MIC), the Ministry of Economy, Trade and Industry (METI), and the Ministry of Land, Infrastructure, Transport and Tourism (MLIT) implemented these projects with the intention of using the outcomes in both domestic and international contexts. The projects also aimed to establish relevant architecture, standards, and data for the construction of smart cities. The Japanese government established the Smart City Public–Private Partnership Platform as part of its strategy in 2019. The office and ministries also launched various smart city projects then.

The 6th Science, Technology, and Innovation Basic Plan, approved by the Cabinet in 2021, clearly establishes the link between the use of ICTs and environmental sustainability. This integration of environmental considerations into smart city policies is a significant achievement. To ensure data interoperability and system scalability between cities and regions, the Japanese government will accelerate the implementation of city OS (data linkage infrastructure) in each region while referring to the Smart City Reference Architecture. The government has already initiated national support in FY2021 to contribute to the progress of efforts towards zero-carbon cities using a wide variety of big data. This plan defines a zero-carbon city as a local government that will achieve zero CO_2 emissions by 2050. Table 7.2 displays the status of zero-carbon city declarations as of 28 December 2023. The national support for zero-carbon cities has been implemented by the CO, MIC, METI, MLIT, and the Ministry of the Environment (MoE).

The Japanese government has developed guidebooks and guidelines for local governments, businesses, and citizens to implement smart city projects effectively. The Smart City Reference Architecture White Paper was published in Japanese in 2020. Its purpose is to provide a comprehensive and detailed description of the necessary components and their implementation guidelines. This will enable local governments and organisations to understand the overall structure and requirements for building strategies for smart cities. The white paper comprises the following foundational components: the Smart City Strategy consists of a set of Smart City

Table 7.2 The declaration situation of zero-carbon cities (as of 28 December 2023)

	Prefectures	*Cities*	*Special wards of Tokyo*	*Towns*	*Villages*	*Total*
Zero-carbon cities	46	570	22	327	48	1013
Percentage (%)	97.9	73.8	95.7	44.0	25.4	58.8
Total	47	772	23	743	189	1724

Source: MoE (2023)

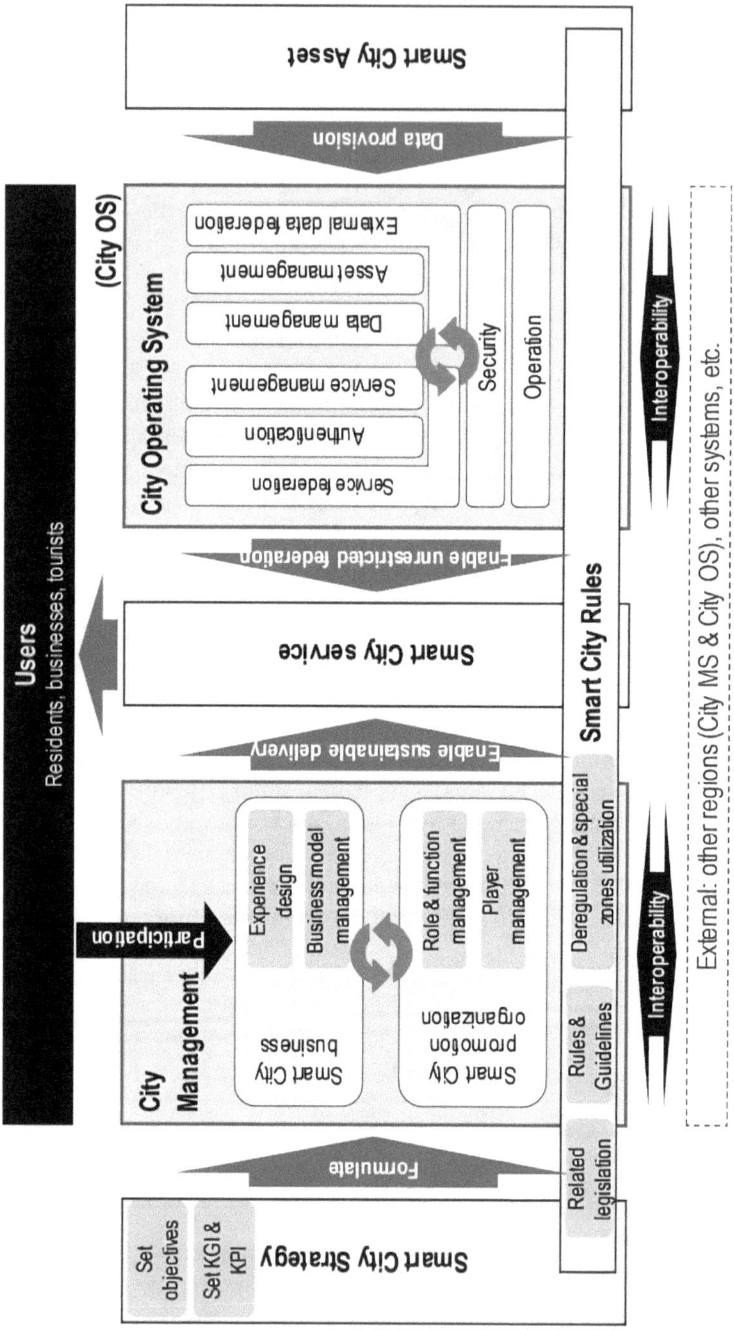

Figure 7.1 Overall picture of Smart City Reference Architecture.

Source: Cabinet Office (2020).

Rules and quantitatively assessable indicators, namely Key Goal Indicators (KGI) and Key Performance Indicators (KPI) (Figure 7.1). The Smart City Rules encompass City Management, Smart City Service, and City OS. This document is referred to as a guidebook. In 2020, the guidebook titled *How to Use Smart City Reference Architecture* was published in Japanese as a deliverable and useful version of this white paper for local governments or entities driving smart city initiatives.

Two years later, in 2022, guidelines for setting KPIs for smart city measures were developed in Japanese to measure the progress achieved after implementing smart city initiatives. These guidelines propose three outcome indicators for mobility-related policies in the environment sector: the car usage rate, indicating the proportion of cars in traffic; the public transport usage rate, addressing the proportion of public transport in traffic; and a person-trip survey. The Municipal Emissions Chart prepared by MoE is used to evaluate the CO_2 emissions from energy sources in the transport sector, which falls under the third category.

In 2020, MLIT published Guidelines for the Linkage of MaaS-related Data, version 1.0, for this purpose. To obtain more accurate human-mobility-related data in the region, including movement from the place of origin to the destination and activities at the destination, it is essential to follow specific guidelines for MaaS projects. These guidelines are the authoritative source for linking MaaS-related data and should be followed to ensure the accuracy of the data. Linking data solves various regional issues, such as restructuring public transport networks, stimulating travel demand, and facilitating effective town planning and infrastructure development. It is also expected to contribute to the revitalisation of local economies and communities.

7.2.2 Characteristics of Smart City Initiatives

There are differences in the projects of smart city initiatives implemented by the CO, MIC, MLIT, and METI (Table 7.3). The CO has successfully implemented the Project for Social Implementation of Future Technologies to establish a local support system (regional implementation council) for the social implementation of future technologies of local governments that excel from the MIC. The MIC's Smart City "Promotion Project for Data Utilisation" and "Smart City Promotion Project for Data Linkage Promotion" create a robust infrastructure platform and data linkage infrastructure, respectively. The Smart City Promotion Project for Data Linkage Promotion will successfully introduce a data linkage infrastructure that guarantees interoperability, scalability, and security, enabling seamless cross-sectoral linkage. The MLIT's Smart City Model Project and the Japanese MaaS Promotion and Support Project will lead the nation in the field of smart cities and develop the infrastructure required to realise MaaS. The latter project will effectively support the development of infrastructure required for the realisation of MaaS. The Regional New MaaS Creation Promotion Project by METI promotes MaaS demonstrations in each region to create advanced and sustainable business models. The MLIT and METI will develop these MaaS projects under the "smart mobility challenge projects" from 2019 onwards, and share information on them.

Table 7.3 Overview of smart city initiatives

Names of smart city initiatives	*Initiative overview*
The Ministry of Economy, Trade and Industry (METI) "Project for Creating and Promoting New MaaS in Regions"	In order to implement new mobility services in the region, this project will promote MaaS demonstrations in each region to create advanced and sustainable business models
The Ministry of Land, Infrastructure, Transport and Tourism (MLIT) "Project for Promoting and Supporting Japanese MaaS"	In order to promote MaaS that responds to new needs with/after COVID-19, such as congestion-free travel and personal mobility, this project will not only support such highly public initiatives but will also support the development of the infrastructure necessary for the actual implementation of MaaS and the use of the plan approval and council system newly established by the revision of the law
The Ministry of Land, Infrastructure, Transport and Tourism (MLIT) "Smart City Model Project"	Aiming to become a global leader in smart cities, the project is looking for leading model projects to develop pioneering initiatives that will guide the whole country and support smart city initiatives
The Cabinet Office (CO) "Project for Social Implementation of Future Technologies"	In order to realise Society 5.0 in the region, this project will provide comprehensive support from relevant ministries and agencies, such as the establishment of a local support system (regional implementation council) for the social implementation of future technologies by local governments that excel in terms of regional revitalisation
Ministry of Internal Affairs and Communications (MIC) "Data Utilization-type Smart City Promotion Project"	This project will develop a platform to collect and analyse data from a variety of sources, with the aim of improving the efficiency and sophistication of local functions and services, enhancing the convenience and comfort of daily life, and creating cities where people can live safely and securely
Ministry of Internal Affairs and Communications (MIC) "Data Linkage-type Smart City Promotion Project"	In order to address various issues facing the region, this project will promote the realisation of a smart city where a variety of services can be provided by promoting the implementation of a data connectivity platform that ensures interoperability, scalability, and security, enabling cross-sector collaboration

Source: The contents in this table are derived from various home pages from the websites of METI, MLIT, CO, and MIC.

Table 7.4 The number of smart city projects by each smart city initiative (as of 31 March 2023)

Names of smart city initiatives	*Number of projects*	*Percent (%)*
METI "Project for Creating and Promoting New MaaS in Regions"	52	18.3
MLIT "Project for Promoting and Supporting Japanese MaaS"	75	26.4
MLIT "Smart City Model Project"	68	23.9
CO "Project for Social Implementation of Future Technologies"	49	17.3
MIC "Data Utilization-type Smart City Promotion Project" and "Data Linkage-type Smart City Promotion Project"	40	14.1
Total	284	100.0

Source: MLIT (2023).

Notably, the proportion of MaaS projects led by the MLIT and METI is significantly high among all smart city projects (Table 7.4). It should be noted that only a small number of smart city initiatives focus on the environment and energy (Table 7.5). Mobility-related projects are the most common type of smart city initiative, followed by tourism-related projects, and the table confidently asserts that MaaS projects are central to smart city initiatives in Japan. The finding of a large proportion of tourism-related projects is consistent with that of Leung *et al.* (2023). This study aims to determine the impact of MaaS on behavioural change and environmental sustainability through four MaaS projects with environmental and energy objectives.

7.3 Effects of the MaaS Projects

The four projects are summarised in Table 7.6. The table presents the project name, project period, project target areas, entities driving the project, project purposes, project overview, and numerical targets. However, some projects do not have KPIs or numerical targets.

7.3.1 Namie I-DO Project

In 2022, the project was implemented in the town of Namie, Fukushima prefecture, despite the challenges posed by the 2011 Great East Japan Earthquake. The town of Namie, located adjacent to the town of Futaba, where the Fukushima Daiichi Nuclear Power Plant was located, has a resident population of just over 2,000 as of the end of January 2024, with approximately 130,000 people still displaced outside the town. Nearly 80% of the town's area remains a difficult-to-return zone. The project was implemented as part of the reconstruction effort following the Great East Japan Earthquake.

The project in the town of Namie also addressed identifying the mobility demands of the townspeople and creating an efficient transportation plan. The

Table 7.5 MaaS project objectives

Names of smart city initiatives	*Transport mobility*	*Tourism/ regional vitalisation*	*Health/ medical care*	*Natural disaster prevention*	*Infrastructure maintenance*	*Logistics*	*Security*	*Urban planning*	*Energy*	*Environment*	*Agriculture, fishery, and forestry*	*Others*	*Total*
METI "Project for Creating and Promoting New MaaS in Regions"	50 49.0%	21 20.6%	8 7.8%	0 0.0%	15 14.7%	4 3.9%	0 0.0%	3 2.9%	1 1.0%	0 0.0%	0 0.0%	0 0.0%	102 100.0%
MLIT "Project for Promoting and Supporting Japanese MaaS"	75 40.1%	74 39.6%	18 9.6%	3 1.6%	0 0.0%	5 2.7%	6 3.2%	2 1.1%	1 0.5%	2 1.1%	1 0.5%	0 0.0%	187 100.0%
All the smart city initiatives	229 28.5%	205 25.5%	78 9.7%	56 7.0%	44 5.5%	41 5.1%	34 4.2%	32 4.0%	25 3.1%	22 2.7%	20 2.5%	17 2.1%	803 100.0%

Source: MLIT (2023a).

Table 7.6 Overview of the MaaS projects with environment or energy objectives

Project name	*Namie I-DO Project*	*Region-wide carbon-neutral project using tourism-type MaaS "Miura COCOON"*	*Universal MaaS~Towards a world where no one gives up on moving*	*Okinawa Smart Shift project (Social implementation of MaaS in Okinawa prefecture)*
Period	7 October – 22 December 2022	November 2022 – January 2023	3–28 February 2022	February–December 2022
Location	Namie town	Yokosuka city, Miura city, Zushi city, Hayama town, Kamakura city	Yokosuka city	Okinawa prefecture
MaaS level	1	2	1	2
Entities driving the Maas project (**Executive secretary**)	**Zenrin Co., Ltd.**, Namie Town, Chodai Co., Ltd., Ray Frontier Co., Ltd., Namie Town Transportation Operator	Tourism MaaS "Miura COCOON" Promotion Council: **Keikyu Corporation**, Jordan Co., Ltd., Keikyu Ad Enterprise Co., Ltd., Miura city, Yokosuka city, Zushi city, Hayama town, Kamakura city, Kanagawa prefecture, etc.	**All Nippon Airways**, Keikyu Corporation, Yokosuka city, Yokohama National University, To Japan Railway Company, MK Taxi, Sompo Japan Insurance Co., Ltd., Prime Assistance, Ashirase, ANA Wingfellows Vui Prince, etc.	Okinawa Smart Shift Consortium: **Daiichi Kotsu Sangyo**, Motobu town, Okinawa Daiichi Kotsu Group, Okinawa Toyota Group, Sompo Japan Insurance, NTT DoCoMo Kyushu Branch, Okinawa prefecture, Naha city, Naha City Tourism Association, etc.
Purposes	In order to realise the social implementation of MaaS in Namie Town, the mobility needs of the town's residents will be assessed and an efficient transport plan will be created	• Encourage behavioural change to shift to public transport through MaaS • Visualise GHG emissions through MaaS and promote its use in conjunction with digital ticketing, thereby reducing traffic congestion and GHG emissions	From the point of view of users, local authorities, communities, and businesses, it solves the problems of those who are reluctant to move and stimulates new mobility demand by encouraging behavioural change	• Elimination of traffic congestion on National Highway 58, a major artery, and distribution of traffic to different modes of transport • MaaS will be used to help each target municipality achieve its urban planning and tourism goals

(Continued)

Table 7.6 (Continued)

Project name	*Namie I-DO Project*	*Region-wide carbon-neutral project using tourism-type MaaS "Miura COCOON"*	*Universal MaaS~Towards a world where no one gives up on moving*	*Okinawa Smart Shift project (Social implementation of MaaS in Okinawa prefecture)*
				• Improve convenience, realise seamless travel, and add experience by creating time • Build transport with a view to the post-Corona period, such as the increase in tourists • Reducing CO_2 emissions through modal shift and the use of electric vehicles • Revitalisation of local economies through the use of mobile data
Project overview	By using a smartphone app with a point service for movement and an information distribution function, a system is created that can increase the motivation of users, and users' movement data is continuously collected by returning points as they move	In addition to strengthening the functions of the tourism-type MaaS "Miura COCOON", which will be deployed as a permanent service from FY2020, and implementing a "visualisation function for the effect of reducing greenhouse gas emissions", a sustainable model will be built by aiming to solve social issues towards achieving carbon neutrality in the entire region by implementing preferential measures for visitors using public transport	The problems faced by people who are reluctant to move for whatever reason will be solved from both a user and business perspective with the concept of Universal MaaS, which adds MaaS to Universal Design, and new mobility demand will be stimulated by encouraging behavioural change	The project aims to achieve seamless mobility and disperse traffic by connecting different modes of transport with the proven MaaS app "my route" and improving access to secondary transport through AI on-demand traffic driving, solving local issues such as chronic traffic congestion, dispersion to public transport, and carbon neutrality

Key performace indicators (KPIs)	(1) Travel distances (2) The number of travels (3) The number of times public transportation is used	(1) Number of site visits (2) Number of registered members (3) Number of digital tickets sold (4) GHG emission reductions (5) Number of activity reservations (6) Overall satisfaction with the entire MaaS service (7) Percentage of people whose tourism behaviour changed as a result of the services provided (8) Percentage of people who chose public transportation as a result of this MaaS	(1) User satisfaction with the services provided (2) Probability of moving according to the route search service route (+user's buffer time) (3) Destination arrival rate of autonomous mobility support services and functions	(1) Number of site visits (2) Rates of each mobility mode (3) Number of digital tickets sold
Numerical targets	(1) Travel distance: 10% increase (2) The number of travels: 10% increase (3) The number of times public transportation is used: 10% increase	(Unknown)	(1) User satisfaction with the services provided: 75% or more of users answered in the top 2 out of 5 (2) Probability of moving according to the route search service (+user buffer time): 75% or more arrive on time (3) Destination arrival rate of autonomous mobility support services and functions: 75% or more arrive on time	(Unknown)

Source: The contents in this table are derived from various reports issued by METI and MLIT.

introduction of MaaS allowed for the modification of the townspeople's behaviour by measuring their behavioural records through the MaaS application installed on their smartphones. Points were awarded to users who provided data, which could be exchanged for various rewards such as bathing tickets at public bathing facilities in Namie town and drinking water at shops run by the general association. The MaaS application discussed here is limited to providing information and storing data so that it is classified as a MaaS Level 1, as it does not facilitate transport reservations or payments. The project aimed to achieve a 10% increase in public transport use as one of its three KPIs. As a result, the desired outcome was a change in behaviour towards public transport use.

7.3.2 Miura COCOON Project

Miura, about 60 km south of central Tokyo, has a population of about 40,000. It is a popular year-round tourist destination for visitors from the Tokyo metropolitan area due to its mild climate and scenic location. Miura is also the site of a project aimed at encouraging a shift towards public transport through MaaS and reducing traffic congestion and GHG emissions.

To achieve this, the private railway company Keikyu Corporation developed the MaaS app "COCOON", which enables users to book and pay for digital tickets. The project's performance was measured using eight carefully selected indicators. Two indicators, namely indicators no. 7 and no. 8, measure behaviour change resulting from the MaaS services provided, while an indicator, namely no. 4, measures the environmental sustainability impact of a MaaS project through GHG emission reductions. No official numerical targets were set for these indicators, but their impact was assessed on a voluntary basis.

7.3.3 Universal MaaS Project

The city of Yokosuka, located about 45 km south of central Tokyo in Kanagawa prefecture, is an industrial and port city with a population of approximately 370,000. It is conveniently located near Haneda Airport, with the shortest journey taking just over an hour. The project provides practical solutions for those who are hesitant to travel, even if public transportation requires multiple transfers.

The project aimed to provide solutions for individuals who are reluctant to move. This was done by considering the viewpoints of MaaS app users, local authorities, communities, and businesses. The ultimate goal was to encourage behavioural change and generate new mobility demand, with a specific focus on promoting air travel through the introduction of a MaaS app. All Nippon Airways (ANA) has developed an MaaS app called "Airport Access Navi" (currently named "Tabi CUBE") that enables users to conveniently access the airport using the airport access Navi system. The app empowers users to search, book, and pay for transportation to their destination with ease. This indicates that the MaaS app has been rated at Level 2. While no outcome indicators were set to directly express the change in behaviour of MaaS users.

7.3.4 Okinawa Smart Shift Project

Okinawa prefecture, located in the southernmost part of the Japanese archipelago, with a population of just under 1,500,000, is a highly regarded year-round tourist destination in Japan due to its mild climate and unique Ryukyuan culture. However, heavy congestion on the main roads in the main tourist areas during peak tourist season is a persistent issue, as there is no railway as a public transport system.

The project aimed to reduce traffic congestion on National Highway 58 and lower GHG emissions through modal and EV shift. The project was intended to have a positive impact on the environmental sustainability of MaaS by achieving these goals. The MaaS application used in the project was "My Route", developed by Toyota Motor Corporation. My Route enables users to book and pay for digital tickets, which is why the level of MaaS is rated at 2. To measure modal shift, a KPI was presented, specifically the proportion of each mobility mode. The project objective of reducing GHG emissions was not assessed using any indicators, and no KPI target was set.

7.3.5 Characteristics of Four MaaS Projects

Four MaaS projects with environmental and energy purposes are characterised as follows. First, only MaaS level 1 or 2 are currently available. However, after the implementation of these four projects, integration with policies or plans is being sought, although not up to the policy integration targeted in MaaS Level 4. It is worth noting that this situation is similar to that in other developed countries (Sochor *et al.*, 2018).

Second, the Namie I-DO, the Miura COCOON, and the Universal MaaS projects aimed to alter the behaviour of MaaS users towards public transport use. However, the ultimate goal of behaviour change was not directly linked to environmental sustainability, such as the reduction of GHG emissions.

Third, in contrast, the Miura COCOON and the Okinawa Smart Shift projects confidently consider the positive impact of MaaS on environmental sustainability. The projects aimed to reduce traffic congestion caused by visitors to the project area by promoting the use of public transport. The main objective is not to shift residents in the project area from private cars to public transport. The reason for the lack of modal shift in the project area may be due to the fact that the Miura COCOON and the Okinawa Smart Shift projects have a focus on tourism rather than mobility.

The impact analysis excluded the Universal MaaS and Okinawa Smart Shift project due to the lack of specific outcomes presented. Instead, the behavioural changes of MaaS users towards public transport were identified through the outcomes of the Namie I-DO and Miura COCOON projects. Only the results of the Miura COCOON project will be used to determine the impact of the MaaS project on environmental sustainability.

Tables 7.7 and 7.8 summarise the results of the Namie I-DO Project and the Miura COCOON project, respectively.

Table 7.7 Outcomes of the Namie I-DO project

Key performance indicators (KPIs)	*No points period*	*Period with points (triple points)*	*Percentage changed (%)*
(1) Travel distances (km/day·person)	9.23	9.10	–1.4
(2) The number of travels (times/day·person)	4.91	5.02	2.2
(3) The number of times public transportation is used (times/week)	3	8	166.7
The number of times specific transport mode is used (times/week)			
Transport mode	*No points period*	*Period with points (triple points)*	*Percentage changed (%)*
Public transportation	3	8	166.7
Private cars	743	770	3.6
Walking	946	1,159	22.5
Bicycle	52	42	–19.2

Source: METI (2023).

Table 7.8 Outcomes of the Miura COCOON project

Key performance indicators (KPIs)	*Before the MaaS project (as of 30 October 2022)*	*After the MaaS project (as of 30 January 2023)*	*Percentage changed (%)*	*Self-evaluation*
(1) Number of site visits	1,50,000	1,74,795	16.5	Fulfilled
(2) Number of registered members	5,000	8,566	71.3	Fulfilled
(3) Number of digital tickets sold	10,000	8,405	–16.0	Unfulfilled
(4) GHG emission reductions (t)	50	48	–4.0	Unfulfilled
(5) Number of activity reservations	3,000	1,103	–63.2	Unfulfilled
(6) Overall satisfaction with the entire MaaS service	80%	56%	–24.0	Unfulfilled
(7) Percentage of people whose tourism behaviour changed as a result of the services provided	70%	80%	10.0	Fulfilled
(8) Percentage of people who chose public transportation as a result of this MaaS	30%	84%	54.0	Fulfilled

Source: MLIT (2023b).

The results provide strong evidence for the effectiveness of MaaS projects in promoting the use of public transport. This is demonstrated by the significant increase (+166.7%) in the use of public transport in the Namie I-DO project and the increase (+54.0%) in the Miura COCOON project. It can be concluded that MaaS projects effectively promote the use of public transport.

The Namie I-DO project did not result in a modal shift from private cars to public transport, as private car use increased slightly by 3.6%. However, this outcome was expected due to the nature of the project and the current transportation infrastructure. It is evident from the fact-finding obtained from a single case study that a MaaS project alone may not be sufficient to cause a modal shift.

Finally, the Miura COCOON project was assessed for its impact on environmental sustainability using the KPI for GHG emissions reduction. The KPI for GHG emission increased by 4.0% in this project, indicating a negative impact on environmental sustainability. However, it is important to note that this conclusion was based on a single case study.

7.4 Discussion

This section highlights the behavioural changes of MaaS users towards the use of public transport and the impact of a MaaS project on environmental sustainability. The author stresses the crucial role of attitude change in the Namie I-DO project, which is expected to bring about significant behavioural changes. This project incentivised users with points to encourage their participation and engagement. The author confidently discusses the environmental sustainability impacts of MaaS projects, with a specific focus on MaaS Level 3, which involves subscription or bundling packages.

Ajzen's (1991) model asserts that behaviour is determined by intention, which is influenced by three factors: attitude towards the behaviour, subjective norm, and perceived behavioural control. Therefore, the use of public transport through MaaS apps is influenced by the attitude that using public transport will enhance environmental sustainability. Moreover, financial incentives, such as points awarded for using public transport, are likely to strongly encourage a positive attitude towards it.

Research on user attitudes towards MaaS is currently underway (Alonso-González *et al.*, 2020; Matowicki *et al.*, 2022). The impact of point systems or special offers on MaaS usage has not been extensively researched. Kamargianni *et al.* (2018) found that while 36% of respondents agreed that MaaS special offers would motivate them to subscribe to MaaS, 41% disagreed. This suggests that the financial incentive of special offers did not necessarily shape positive attitudes towards MaaS. However, the Namie I-DO project found that the use of public transport increased due to the addition of points. Limited impact of attitudes on behaviour may result in contradictions regarding point systems or special offers (Schikofsky, Dannewald, and Kowald, 2020). Therefore, further research is required to determine the impact of attitude change on behaviour.

A subscription or bundling package of MaaS could enhance its positive impacts on environmental sustainability. The Miura COCOON project implemented a MaaS

Level 2 system for tourists visiting the Miura Peninsula region, which also allowed visitors outside the area to use the system. For such reasons, the Miura COCOON project increased public transport use but did not reduce GHG emissions. Using a MaaS Level 3 system, based on a subscription or bundling package for residents within the Miura Peninsula region, and limiting use of the MaaS level 3 app to such residents, will increase public transport use and reduce the number of private cars. As a result, GHG emissions in the region will decrease. However, previous studies (Caiati *et al.*, 2020; Guidon *et al.*, 2020; Kriswardhana and Esztergár-Kiss, 2023) had not established a clear relationship between MaaS subscriptions or bundling packages and GHG emission reduction. With further research, therefore, the impact of these packages on GHG emissions can be better understood.

In addition to promoting environmental sustainability, the introduction of MaaS Level 3 could lead to energy savings. By encouraging a modal shift from private cars to public transport, the implementation of MaaS Level 3 could reduce GHG emissions as a primary effect, with energy savings as a secondary effect. A scenario analysis conducted in Sweden demonstrated that the implementation of MaaS would result in a substantial reduction in car usage, leading to lower energy consumption, especially when public transport and other mobile services are utilised more (Zhao *et al.*, 2021). In contrast, a case study conducted in Zurich showed that MaaS significantly reduced energy consumption in the passenger sector (Becker *et al.*, 2020). Furthermore, MaaS is being used to integrate parcel delivery using electric vehicles, which is also expected to result in energy savings in freight transport (He and Csiszár, 2021). These studies suggest that MaaS can facilitate energy transitions. Therefore, the use of ICTs and IoTs in the construction of smart cities is expected to enhance energy transitions, as MaaS can be considered a derivative of smart cities (Leonhardt *et al.*, 2024; Sokołowski and Visvizi, 2023).

7.5 Conclusion

As identified in this chapter, there are changes in user behaviours resulting from MaaS and its impact on environmental sustainability. The empirical analysis, which focused on a MaaS project implemented by the Japanese government, clearly demonstrated behavioural changes in MaaS users towards the use of public transport. While it was not possible to definitively conclude that a modal shift from private vehicles to public transport has occurred, the evidence presented strongly suggests that such a shift is underway. The research findings suggested that the impact of MaaS-related projects on environmental sustainability was not clearly demonstrated due to the operating conditions of MaaS that rely on MaaS level 1 or 2. The modal shift and its impact on environmental sustainability lacked visibility as a result. These findings demonstrated the need for further research and development in this area to fully understand the potential impact of MaaS on environmental sustainability. The policy implications from MaaS Level 2 towards Level 4, which aims at policy integration, are described in the following part.

The policy implications for environmental sustainability are described in terms of dynamic location information, the development of a platform for the effective use of such data, and policy evaluation after policy implementation. First, with regard to the handling of dynamic location information, recent mobility policies need to focus on CASE, the connectivity of which may not be adequately addressed in future MaaS policies under the existing Japanese legal system. The term "CASE" was coined by Dr Dieter Zetsche, the former chairman of Mercedes-Benz, during a press conference at the Paris Motor Show in 2016. The term stands for connectivity to the internet, autonomous driving, shared vehicles, and electric drive systems. In MaaS Level 4, therefore, each vehicle is connected to the internet, and the real-time location information acquired by each vehicle is stored and analysed via the internet on the city platform consisting of the city OS (Ruohomaa and Salminen, 2019). As a result, real-time information such as traffic jam avoidance and the shortest route is presented to MaaS users, which is expected to reduce GHG emissions in the transport sector.

In anticipation of this situation, the European Union's revised ITS Directive, which was enacted in 2023, lists the following two prerequisites for supporting the development of accurate mobility management services provided by public transport authorities, including MaaS. The Directive (2023) also includes the following two conditions to support the development of accurate mobility management services provided by public transport authorities, including MaaS. Namely, the "availability and accessibility of existing, accurate road and multimodal travel and traffic data needed for mobility management, in a standardised format, to the relevant public authorities without prejudice to data protection requirements" (2.2.1). "The timely updating of available road and multimodal travel and traffic data needed for mobility management by the relevant public authorities and stakeholders" (2.2.3).

In Japan, under the Act on the Protection of Personal Information, which came into force in April 2022, location information that may identify an individual by being continuously stored or otherwise is considered "personal information", but location information of an individual that cannot identify a specific individual is not.[1] However, location information of an individual that cannot identify a specific individual falls under "location-related information" and can be used and provided by private companies and the general public if appropriate measures are taken. Meanwhile, Article 6 of the Basic Act on the Advancement of Public and Private Sector Data Utilisation,[2] enacted in 2016, sets out the responsibilities of private sector operators, stating that operators shall comply with the Basic Act on the Advancement of Public and Private Sector Data Utilisation. This article also states that private business operators shall, in accordance with the basic act, actively promote the use of public and private sector data in their business activities and cooperate with measures taken by the central or local authorities to promote the use of public and private sector data. Based on this article, the provision of location information obtained by private companies through mobile terminals owned by individual vehicles and the general public can only be done at a cost. Therefore, such location information may not always be provided to MaaS in a timely manner.

In light of this, although the government has already provided guidelines on the security of smart cities and the quality of data provided through data infrastructure, guidelines on the active provision of data, including open data, and a gentleman's agreement on the provision of data in the public and private sectors are considered necessary.

Second, with regard to the development of infrastructure for the effective use of data, a city platform is necessary when aiming for MaaS Level 4, but other smart city projects being implemented by the CO and MIC might not be tightly linked to the MaaS projects. As mentioned earlier, the effective use of location information through the operation of a city platform based on a city OS such as FIWARE is essential to ensure the effectiveness of MaaS Level 4 (Al-Rahamneh *et al.*, 2021; Yamashita, 2022). However, the development of city OSs has mainly been promoted by the MIC through the Data Utilisation Type Smart City Promotion Project and the Data Linking Type Smart City Promotion Project. Although the knowledge, information, etc. accumulated in all the smart city projects have been shared through the Smart City Public–Private Partnership Platform, the establishment of a city platform requires high costs, so it was desirable to develop MaaS projects in the municipalities implementing the two data-related projects by the MIC. In order to ensure the effectiveness of MaaS Level 4 in the future, it is also considered necessary to overlap city platform development projects with MaaS projects. This would also help to reduce costs for both central and local governments.

Finally, policy evaluation needs to be strengthened to articulate the positive impacts of MaaS projects on environmental sustainability. As mentioned earlier, the Japanese government provided guidance on KPIs for smart city initiatives. These KPIs included indicators to measure environmental sustainability, such as the reduction of GHG emissions. However, as seen in the Okinawa Smart Shift project, despite the project's stated goal of GHG emission reductions, the KPIs did not include GHG emission reductions, which meant that the environmental sustainability impact of the project could not be measured. Therefore, future MaaS projects are encouraged to establish more detailed guidelines for the use of KPIs, such as presenting mandatory indicators from the KPI guidelines if their objectives include an environmental dimension.

Acknowledgements

This research was partly supported by JSPS KAKENHI (grant number 21K01030).

Notes

1 In Japanese: 個人情報の保護に関する法律 [kojin joho no hogo ni kansuru houritsu] also known as 個人情報保護法[kojin joho hogo hou].

2 In Japanese: 官民データ活用推進基本法[kanmin data katsuyou suishin kihon hou].

References

Ajzen, I. (1991) "The Theory of Planned Behaviour", *Organizational Behavior and Human Decision Processes*, 50(2), pp. 179–211.

Alonso-González, M.J. *et al.* (2020) "Drivers and Barriers in Adopting Mobility as a Service (MaaS) – A Latent Class Cluster Analysis of Attitudes", *Transportation Research Part A: Policy and Practice*, 132, pp. 378–401.

Al-Rahamneh, A. *et al.* (2021) "Enabling Customizable Services for Multimodal Smart Mobility with City-Platforms", *IEEE Access*, 9, pp. 41628–41646.

Arias-Molinares, D. and García-Palomares, J.C. (2020) "The Ws of MaaS: Understanding Mobility as a Service from a Literature Review", *IATSS Research*, 44(3), pp. 253–263.

Banister, D. and Stead, D. (2004) "Impact of Information and Communications Technology on Transport", *Transport Reviews*, 24(5), pp. 611–632.

Becker, H. *et al.* (2020) "Assessing the Welfare Impacts of Shared Mobility and Mobility as a Service (MaaS)", *Transportation Research Part A: Policy and Practice*, 131, pp. 228–243.

Cabinet Office (2020) "Smart City Reference Architecture White Paper (First Edition)". Available at: https://www8.cao.go.jp/cstp/stmain/a-whitepaper1_200331_en.pdf(Accessed: 18 February 2024).

Caiati, V., Rasouli, S. and Timmermans, H. (2020) "Bundling, Pricing Schemes and Extra Features Preferences for Mobility as a Service: Sequential Portfolio Choice Experiment", *Transportation Research Part A: Policy and Practice*, 131, pp. 123–148.

Chawla, Y., Shimpo, F. and Sokołowski, M.M. (2022) "Artificial Intelligence and Information Management in the Energy Transition of India: Lessons from the Global IT Heart", *Digital Policy, Regulation and Governance*, 24(1), pp. 17–29.

D'Amico, A. (2023) "New Frontiers for Sustainable Mobility: MaaS (Mobility as a Service)", *TeMA – Journal of Land Use, Mobility and Environment*, 16(2), pp. 455–460.

Directive (EU) 2023/2661 (2023) Directive of the European Parliament and of the Council of 22 November 2023 Amending Directive 2010/40/EU on the Framework for the Deployment of Intelligent Transport Systems in the Field of Road Transport and for Interfaces with Other Modes of Transport, O.J. L 2023/2661.

Directive 2010/40/EU (2010) The European Parliament and of the Council of 7 July 2010 on the Framework for the Deployment of Intelligent Transport Systems in the field of Road Transport and for Interfaces with Other Modes of Transport. O.J. L 207.

Guidon, S. *et al.* (2020) "Transportation Service Bundling – For Whose Benefit? Consumer Valuation of Pure Bundling in The Passenger Transportation Market", *Transportation Research Part A: Policy and Practice*, 131, pp. 91–106.

He, Y. and Csiszár, C. (2021) "Model for Crowdsourced Parcel Delivery Embedded into Mobility as a Service Based on Autonomous Electric Vehicles", *Energies,* 14(11), 3042.

Hensher, D.A. *et al.* (2020) *Understanding Mobility as a Service (MaaS): Past, Present and Future*. Amsterdam: Elsevier.

Hensher, D.A., Ho, C.Q. and Reck, D.J. (2021) "Mobility as a Service and Private Car Use: Evidence from the Sydney MaaS Trial", *Transportation Research Part A: Policy and Practice*, 145, pp. 17–33.

Hesselgren, M., Sjöman, M. and Pernestål, A. (2020) "Understanding User Practices in Mobility Service Systems: Results From Studying Large Scale Corporate MaaS in Practice", *Travel Behaviour and Society*, 21, pp. 318–327.

Hörcher, D. and Graham, D.J. (2020) "MaaS Economics: Should We Fight Car Ownership with Subscriptions to Alternative Modes?", *Economics of Transportation*, 22, 100167.

Kamargianni, M. *et al.* (2016) "A Critical Review of New Mobility Services for Urban Transport", *Transportation Research Procedia*, 14, pp. 3294–3303.

Kamargianni, M. *et al.* (2018) *Londoners' Attitudes towards Car-Ownership and Mobility-as-a-Service: Impact Assessment and Opportunities That Lie Ahead*. London: MaaSLab – UCL Energy Institute for Transport for London.

Köhler, J. *et al.* (2019) "An Agenda for Sustainability Transitions Research: State of the Art and Future Directions", *Environmental Innovation and Societal Transitions*, 31, pp. 1–32.

Kriswardhana, W. and Esztergár-Kiss, D. (2023) "A Systematic Literature Review of Mobility as a Service: Examining the Socio-Technical Factors in MaaS Adoption and Bundling Packages", *Travel Behaviour and Society*, 31, pp. 232–243.

Kriukelyte, E. (2018) "New Challenges for Transport Planning: The Institutionalization of Mobility as a service in the Stockholm Region". Available at: www.diva-portal.org/smash/record.jsf?pid=diva2%3A1187411&dswid=-5759 (Accessed: 15 April 2024).

Lauri, C., Shimpo, F. and Sokołowski, M.M. (2023) "Artificial Intelligence and Robotics on the Frontlines of the Pandemic Response: The Regulatory Models for Technology Adoption and the Development of Resilient Organisations in Smart Cities", *Journal of Ambient Intelligence and Humanized Computing*, 14(11), pp. 14753–14764.

Leung, A., Burke, M. and Scott, P. (2023) "Tourism MaaS – The case for regional cities". *Research in Transportation Business and Management*, 49, 101017.

Leonhardt, S. *et al.* (eds) (2024) *Innovations and Challenges of the Energy Transition in Smart City Districts*. Berlin: De Gruyter.

Liu, Y. *et al.* (2017) "Exploring Data Validity in Transportation Systems for Smart Cities", *IEEE Communications Magazine*, 55(5), pp. 26–33.

Lyons, G. (2020) "Walking as a Service – Does it Have Legs?", *Transportation Research Part A: Policy and Practice*, 137, pp. 271–284.

Maas, B. (2022) "Literature Review of Mobility as a Service", *Sustainability*, 14(14), p. 14: 8962.

Matowicki, M. *et al.* (2022) "Understanding the Potential of MaaS – An European Survey on Attitudes", *Travel Behaviour and Society*, 27, pp. 204–215.

METI (2023) "スマートモビリティチャレンジ2022" [Smart Mobility Challenge 2022]. Available at: www.meti.go.jp/shingikai/mono_info_service/smart_mobility_challenge/pdf/20230331_s01.pdf (Accessed: 18 February 2024).

Mladenovic, M. (2021) "Mobility as a Service", in R. Vickerman (ed.) *International Encyclopaedia of Transportation*. Amsterdam: Elsevier, pp. 12–18.

MLIT (2023a) "プロジェクト一覧" [Project List]. Available at: www-mlit-go-jp.translate.goog/scpf/projects/docs/projects.xlsx (Accessed: 18 February 2024).

MLIT (2023b) "令和４年度　新モビリティサービス推進事業一覧"" [List of New Mobility Service Promotion Projects for FY2022]. Available at: wwwtb.mlit.go.jp/kanto/content/000294453.pdf (Accessed: 18 February 2024).

MoE (2023) "地方公共団体における2050年二酸化炭素排出実質ゼロ表明の状況" [Status of Declarations for Achieving Net Zero CO2 Emissions by 2050 by Local Governments]. Available at: www.env.go.jp/policy/zerocarbon.html (Accessed: 18 February 2024).

Pangbourne, K. *et al.* (2020) "Questioning Mobility as a Service: Unanticipated Implications for Society and Governance", *Transportation Research Part A: Policy and Practice*, 131, pp. 35–49.

Rey-Moreno, M., Periáñez-Cristóbal, R. and Calvo-Mora, A. (2023) "Reflections on Sustainable Urban Mobility, Mobility as a Service (MaaS) and Adoption Models", *International Journal of Environmental Research and Public Health*, 20(1), 274.

Ruohomaa, H.J. and Salminen, V.K. (2019) "Mobility as a Service in Smart Cities – New Concept for Smart Mobility in Industry 4.0 Framework" in *ISPIM Conference Proceedings*, pp. 1–12.

Santos, G. and Nikolaev, N. (2021) "Mobility as a Service and Public Transport: A Rapid Literature Review and the Case of Moovit", *Sustainability*, 13(7), 3666.

Schikofsky, J., Dannewald, T. and Kowald, M. (2020) "Exploring Motivational Mechanisms Behind the Intention to Adopt Mobility as a Service (MaaS): Insights from Germany", *Transportation Research Part A: Policy and Practice*, 131, pp. 296–312.

Shimpo, F. (2018) "The Principal Japanese AI and Robot Law. Strategy and Research toward Establishing Basic Principles", *Journal of Law and Information System*, 3, pp. 44–65.

Smith, G., Sochor, J. and Sarasini, S. (2018) "Mobility as a Service: Comparing Developments in Sweden and Finland", *Research in Transportation Business & Management*, 27, pp. 36–45.

Sochor, J. *et al.* (2018) "A Topological Approach to Mobility as a Service: A Proposed Tool for Understanding Requirements and Effects, and for Aiding the Integration of Societal Goals", *Research in Transportation Business & Management*, 27(2), pp. 3–14.

Sokołowski, M.M. and Visvizi, A. (eds) (2023) *Routledge Handbook of Energy Communities and Smart Cities*. London: Routledge.

Wittstock, R. and Teuteberg, F. (2019) "Sustainability Impacts of Mobility as a Service: A Scoping Study for Technology Assessment", in F. Teuteberg, M. Hempel and L. Schebek (eds) *Progress in Life Cycle Assessment 2018. Sustainable Production, Life Cycle Engineering and Management.* Cham: Springer, pp. 61–74.

Yamashita, J. (2022) "Smart City Initiatives in Japan: Achievements and Remaining Issues" in R.K. Mishra *et al.* (eds) *Smart Cities for Sustainable Development. Advances in Geographical and Environmental Sciences.* Singapore: Springer, pp. 79–95.

Zhao, X. *et al.* (2021) "Potential Values of Maas Impacts in Future Scenarios", *Journal of Urban Mobility*, 1, p.100005.

8 Central and Local Government Participation in Decentralised Smart City Development Tools as a Manifestation of Modern Methods of Administrative Tasks

Jacek Piecha, Justyna Kamila Kanas, and Piotr Mikusek

8.1 Introduction

Throughout the decades, public administration has undergone a significant transformation, including receiving increasingly more roles and obligations. Public entities are nowadays not simply executives in implementing policies but also directly responsible for providing for citizens' needs (Basu, 2004, pp. 1–2). Not only the functional aspect has been broadened. Likewise, the subjective scope of administration activities is under continuous modification due to civilisation's progress (Basu, 2004, p. 2). The dynamic nature of contemporary societies leads to higher expectations for state bodies to act. This adjustment to a rapidly changing world is one of the administration's biggest challenges, on the other hand, it is deemed as one of its duties (Vocca, 2020, p. 216). Adapting and responding effectively, sometimes beyond traditional administrative features, to the evolving and tangled landscape is an integral part of their mission. In order to fully exploit the potential of globalisation for miscellaneous stakeholders' interest groups, the delivery of such a task needs to be designed carefully (Rondinelli *et al.*, 2008, p. 1). One of the comprehensive solutions to conjoin society's demands with local government activities is a smart city concept (Meijer and Bolívar, 2016, p. 393) on which this chapter is concentrated.

The evolving nature of the smart city concept offers promising prospects for enhancing urban living and addressing societal challenges. Due to the wide range of needs and opportunities provided by the smart city, its definition remains ambiguous. Nevertheless, it is valuable to highlight the aspects of "smartness" concerning local communities. Oleg Dashkevych and Boris A. Portnov (2022) are among those who submitted empirical criteria used for smart city identification covered by different scientific fields. Their concept of smart cities can be put into a three-tier structure, including (i) economy and technology; (ii) environment; and (iii) society, which serve as a starting point in the chapter's background. Three similar key characteristics are commonly accepted (Pianezzi, Mori and Uddin, 2023,

DOI: 10.4324/9781003471417-8

p. 633; Appio, Limab and Sotirios, 2019, p. 7; Aceleanu *et al.*, 2019, p. 1, p. 182). Clearly, it shows the concept of smart cities as multidimensional, encompassing a diverse collection of smaller features. Despite the broad scope of services under the smart cities, the essence shall revolve around the purpose of smart cities' establishment, which increases the residents' quality of life (Dashkevych and Portnov, 2022; Appio, Limab and Sotirios, 2019, p. 2). From this perspective, the public administration, primarily affected by the traditional ruling, currently discovers itself as a co-manager to the intricate web of interconnected systems that shape a smart city with the importance of collaboration with the private sector, academia, and other active members of the local community. Administration entities shall not only understand innovative products brought to everyday life but also implement them in daily usage to be freely accessed by anyone interested. Creating an ecosystem where stakeholders of all sectors cooperate for technological advancement and ensuring smart city initiatives, as well as social inclusion or environment sustainability seems like a horrendous challenge with intensely satisfying results.

Japan is one of the countries swiftly adapting to the constant changes and acknowledging perfectly the smart cities' purpose (Su, Miao and Wang, 2022). Not only does its capital city – Tokyo – implement smart solutions, other municipalities, such as Osaka, Yokohama, and Kyoto (Fietkiewicz and Stock, 2015), also present this approach. It is due to the demand to follow the pace of global development, including the course of energy transformation. In this context, smart city solutions in the context of Japan are oriented towards such goals as rational energy usage or utilisation of heat and unused energy sources (Clarisse, 2014, p. 8). The approach is not merely a response to the post-Fukushima challenges in Japan's energy mix, but nonetheless a strategic key to building the resilience and sustainability of the future of the citizens (Sokołowski and Kurokawa, 2022).

The preceding depiction encourages an extremely intriguing smart city case study on the example of Japan, which this chapter carries out. Within this narrative, the chapter tackles Japan's public administration's innovative approaches to progress of civilisation, showcasing the development and management of smart cities. But before describing the practical approach of authorities to the creation of smart cities, the latest trends affecting the functioning of the administration need to be illustrated. This is because smart cities cannot be treated as a phenomenon isolated from the sphere of broadly understood state administration. The transition from the traditional model of creating cities to the smart cities model is not a process based solely on technological development. It can be analysed from the point of view of the transformations taking place in the methods of administering the state and, more specifically, the new trends in the method of legal administrative regulation.

Distinguishing contemporary trends in the evolution of the functioning of administration, three main phenomena can be noted: deprivatisation (widening of the areas of life covered by administration), decentralisation (devolution of more powers to the local level), and deprioritisation (a more powerful form of action). Observing these phenomena, one can see that they are the very foundations of smart cities. The early Japanese concepts of smart city development emerged from the second half of the 20th century and the manifestations of the evolving trends became evident with the development of administration during the 21st century. After that, the practical execution of the current demand for public administration activities in the creation of a sustainable economy from smart cities shall be presented. The third section focuses on case studies from municipal sectors, which the literature recognises as strategic in urban development. In these industries, owing to their importance, governmental entities mostly conduct ongoing management. Hence, this formed the rationale for examining municipal sectors in the context of Japan's strategy and endeavours towards establishing a sustainable economy, which also stresses the need for self-sufficiency.

8.2 Smart Cities as an Evolution towards a New Model of Government Functioning

As already indicated, in all social systems and legal orders, public administration is on the frontline and plays a key role as a stakeholder to solve the day-to-day problems of the functioning of modern societies (Lodge and Wegrich, 2014, p. 4). This is particularly evident in crisis situations, where a sudden event, such as a natural disaster affecting the performance of the executive tasks of the state (Kusumasari and Alam, 2012), an epidemic hindering the provision of an adequate level of public services (Špaček, Navrátil and Špalková, 2023), or an armed conflict leading to the activation of management under a long-term "stress test" (Bland, 1942), leads to the need to implement extra-coordinated administrative measures to ensure the functioning of the state in a given area.

At the current juncture, there can be little doubt that climate change and its consequences represent one of the greatest challenges that humanity – and inevitably the administration – will have to deal with (Steffan *et al.*, 2015; Wang and Chen, 2013). As the World Economic Forum (2024, p. 11) points out, as many as four of the top five long-term global risks are related to the environment and climate change. The scale and complexity of the problem justifies the claim that we are dealing with a form of staggering crisis. Indeed, defining a crisis as a situation involving individuals, groups, or entire communities that cannot be dealt with using normal procedures and which occur due to a sudden change (Moe and Pathranarakul, 2006, p. 402), there is no doubt that one of the key stakeholders who will take action on the climate crisis will be the administration.

A traditional approach to defining administration based on Max Weber's notion of "bureaucracy" is as a system for the effective management of the state (Fry, 1989, p. 31). The traditional approach to administration reflected in how "administrative law" in civil law countries is understood emphasises the organisational and managerial character of the activities undertaken by administrations, in which the administration acts as an unequal (stronger, acting as managing bodies) actor in relation to actors outside the administration (weaker, acting as managed bodies) and thus has an imperious character (Piecha 2022, p. 271; Pound, 1941; Ospanova *et al.*, 2018). Increasingly, the literature stresses the need to evolve the classical understanding of "administration" and "administrative law" towards new forms of incidence to meet the challenges of the modern world (Basu, 2019), which can also be applied to the challenges of the changing world, in particular the climate crisis.

The evolution of the understanding of administration from what was previously described as a classic to a modern system of administration can be described by means of three main trends: (i) deprivatisation, consisting of the increasing standardisation of issues hitherto left to the private sphere; (ii) decentralisation and the increasing role of the local level administration being closer to the administered entities; (iii) deprioritisation of the authoritative character of action, consisting the increasing participation of the administration in contacts with entities outside the administration based on non-authoritative forms of action. Deprioritisation is a certain logical consequence of deprivatisation. This is because it makes it possible to intervene more extensively in new spheres of citizens' lives while at the same time satisfying the requirements of proportionality of the law and limiting the administration's interference to methods which are the least intrusive for the administered subjects. All these three main trends are enabled and accelerated by digitalisation and growing social expectations in terms of expanding the catalogue of services provided by the state.

Combining the discussed trends, that is, deprivatisation, decentralisation, and deprioritisation, based on digitalisation, with methods aimed at reducing human impact on the climate, one must conclude that the avant-garde action in this regard is the concept of smart cities, with Japan being one of the first countries to pioneer the implementation of this concept (Kono, Suwa and Ahmad, 2016, p. 97). Japan's public administration bodies started to intensify the smart cities creation in the early 21st century. The impetus was made, among others, by the Ministry of Economy, Trade and Industry (METI), as the principal entity liable for innovation in the industrial area (Yarime, 2020, p. 95), which pinpoints sustainable urban development in the formula of smart cities (Pianezzi, Mori and Uddin, 2023, p. 636).

The advancing development of the concept and implementation of smart city solutions in Japan provides an exceptionally good model for observing the transition from a classical understanding of administration to a modern way of administration. When analysed in the context of the triad of trends modernising administration: deprivatisation, decentralisation, and deprioritisation, this process takes on new meanings and can be part of not only innovation policy but also broadly understood measures to reduce the negative human impact on the climate.

The process of deprivatisation, understood as an increasing number of human activities moving from the strictly private area to the area in the interest of public policies, is a process that is a well-known process in legal theory (Kjaer, 2018). Deprivatisation cannot be understood as simply "nationalising private tasks". It has to be seen in the context of the subject matter (the increase in the number of hitherto unregulated areas and the increase in cooperative tools involving public and private actors). Such seemingly mundane activities from private life as household chores, such as laundry or cooking, and the sphere of home entertainment, such as watching TV, can have a significant impact on the functioning of the energy system due to their specificity, which is also discernible from the Japanese example (Yamaguchi, 2019).

In the context, "deprivatisation" is not a process qualified from the point of view of social philosophy or ethics and does not take on the character of a normative description but a descriptive one. Leaving aside the assessment of the validity of expanding the sphere of interest of public entities (Holstein and Gubrium, 1995), it should be noted that from the point of view of the development of smart cities, the deprivatisation process brings many opportunities in terms of the smart cities concept. As a number of studies indicate, the inclusion of households within the framework of smart cities mechanisms in climate and energy transition processes can bring significant benefits, for example a reduction in decarbonisation costs resulting from the integration of activities into the already ongoing processes of internet of things (IoT) solutions (Paneru and Tarigan, 2023), the spread of pro-efficiency behaviour in energy use (Khansari, Mostashari and Mansouri, 2013, p. 56), or the integration of new solutions into smart cities, thereby creating a more coherent ecosystem that exploits synergies between existing solutions, for example improving the quality of the urban environment through smart waste management (Crome *et al.*, 2023).

However, it would be a mistake to consider the deprivatisation process solely from the point of view of the benefits accruing to the public administration. Such a one-sided flow of benefits would undoubtedly lead to the failure of climate and energy policy due to a fundamental omission of recognition justice, relating to the recognition of the role and needs of individual stakeholders (Heffron and Sokołowski, 2022, p. 7). Administered entities must also have their own individual interest – separate from that of the administration – for participating in smart city solutions (Marrone and Hammerle, 2018) and feeding the system with what is most valuable in it, that is data (Al Nuaimi, 2015; Sta, 2017; Allam and Dhunny, 2019).

There are, however, a number of challenges associated with the operation of smart cities in the context of deprivatisation, a detailed description of which is well beyond the scope of this publication, and it is therefore only necessary to signal the main challenges in this regard. A fundamental issue in this regard is, of course, cyber security (Kitchin and Dodge, 2019, pp. 52–55). It is therefore necessary to provide technological and legal mechanisms to guarantee the highest level of protection for personal data processed within smart cities (Shimpo and Mikusek, 2023). Another issue that represents a significant challenge to be addressed will be the problem of social exclusion resulting from the lack of opportunities for participation within smart cities (Valdez *et al.*, 2021). Actual exclusion may result, for example, from a lack of digital skills (Bleja *et al.*, 2020). Deprivatisation processes bringing more areas of life within the scope of administrative law regulation are taking place all over the world, including Japan (see Chapter 3 for detailed examples).

Another necessary parallel phenomenon sharing modern administration, and the smart cities concept, is the decentralisation processes of public authority. The concept of decentralisation, due to its ambiguity, is a concept that can take on different scopes depending on the objectives of the analysis and the comparative criteria used. It can also take different forms of decentralisation in the form of deconcentration or delegation (Rondinelli, 1990, pp. 9–15). Authors analysing different research approaches to decentralisation in the context of administration highlight the common core of the concept (Cohen and Peterson, 1996; Bardhan, 2002; Schneider, 2003), which can be defined as "(...) the transfer of responsibilities and financial resources from the central government and its bodies" (Cohen and Peterson, 1996, p. 11).

Decentralisation of administration is seen as democratising social life by increasing the number of stakeholders involved in the exercise of power, thereby dispersing power and increasing the number of tasks performed at local levels (Sundar, 2001; Gavkalova and Kolupaieva, 2018, pp. 218–219). It is significant that analogous arguments can be found in the arguments in favour of the development of the smart cities concept. Democratisation through the different stakeholders in these processes, for example based on blockchain technology, is emphasised, where members have an interest in participating for creating a local, more informed user community (Soni and Bhusham, 2019). There is also a noticeable tendency to consider the tools included in smart cities as a certain new field where local communities can exert influence on the administration (Klijn and Koppenjan, 2016; Nesti and Graziano, 2019), which may result in a greater influence on local policies especially on energy and environmental policies.

The parallelism between the development of modern trends in approaches to administration and the concept of smart cities is clearly visible in the example of Japan. As early as 1989, the Japanese Ministry of the Environment prepared the Environmental White Paper, in which the concept of eco-poleis emerged. Eco-poleis were to be environmentally friendly urban systems using modern technology. This concept was continued by the Ministry of Construction launching the environment-symbiotic model city project in 1993 (Okubo *et al.*, 2022, pp. 500–501). Accordingly, to monitor and stimulate possible alliances in this field, in 2003, an institution called the New Energy and Industrial Technology Development Organisation (NEDO) was established in Japan. It promotes the dissemination of industrial, energy, and environmental technologies next to research and development operations (Kakuno, 2014, p. 1). By adopting the legal form of Incorporated Administrative Agency (General Rules for Incorporated Administrative Agencies Act, 1999; New Energy and Industrial Technology Development Organisation Act, 2002), a certain degree of autonomy from the government in its management decision-making is assembled (Hori, 2003), which responds to the need for designing smart solutions by quicker than usual adjustment to emerging challenges and opportunities in urban planning. Also to be superimposed on the processes taking place is the development of e-governance in Japan, that is the digitalisation and transfer of public administration activities to the digital sphere (Purnendra, 2002; Wong, Hideki and George, 2011), which further accelerates and directly links the model of administration development with the development of the smart cities concept. An interesting public policy issue that should be signalled and needs to be investigated separately in the future will be the issue of non-hierarchical approaches within smart cities and the construction of competitive advantages by Japanese industry (Guseva and Dmitrieva, 2020), through cooperation with the local administration. An example of such cooperation can be found in Kitakyushu city, where industry representatives (Nippon Steel Yawata, Nippon Steel Engineering, Mitsui & Co., Ltd., Hibikinada Kaihatsu) formed an advisory committee with city representatives responsible for diverse areas of city operations to support the development of smart projects in the city (Shiroyama and Kajiki, 2016, p. 120).

The example of city of Kitakyushu and the cooperation between industry and the local administration is also a case in point, illustrating the third characteristic of modern public administration, which is inevitably linked to the idea of smart cities, namely the deprioritisation of the sovereign nature of administrative action and the use of legal institutions that strengthen the position of administered entities. Administrative law in its assumptions is based on the exercise of the law in a sovereign manner. The administering subject has the possibility to unilaterally shape the rights and obligations of the administered subjects on the basis and within the limits of the applicable law (Howe, 2002). It is thus a clear reminiscence of the Hobbesian *Leviathan* being an identification of the power of the state based on the legitimacy of law and the actions of the administration as a command (Torgerson, 1990; Hoye, 2019, p. 199).

However, one can clearly discern a trend of "softening" of the aforementioned classical approach of the administrative law-as-command model and the emergence of more and more institutions of increased public participation (Callahan, 2007; Crosby, Kelly and Schaefer, 2015). Analysing the successive phases of the shift from treating public administration as the main and only decision-making centre to inviting the exercise of power and the implementation of public policies onto the stage, three main elements can be distinguished, following Đurman (2020, pp. 86–94): (i) participation in the creation of policies and regulations; (ii) participation in the implementation of policies and regulations, and (iii) participation in the supervision of public administration.

Again, using the chosen method and applying smart cities concepts to the development trends of modern administration, one can see a clear overlap of trends and a mutual symbiosis of possible tools. Japan diversifies its activities, highlighting the cooperation between all interested groups, frequently under a public–private partnership scheme (Pianezzi, Mori and Uddin, 2023). The collaborative framework, involving governmental bodies, local governments, and private counterparts, jointly launches numerous initiatives. According to the model rationale from legal scientific literature (Sokołowski and Visvizi, 2023), it could be classified as energy cooperatives. The implementation of energy co-operatives, which are a tool for operationalising and a means of active citizen participation in smart cities, can be a model way of not empowering citizens in the implementation of public policies. This is because energy co-operatives – in the broadest sense – serve to support the basic energy needs of its members (Gjorgievski, Cundeva and Georghiou, 2021). Thus, they support the local administration in ensuring that the basic needs of the community are met. However, the right shaping through regulation of their position in the system (Sokołowski, 2021) can contribute to the achievement of climate goals (Lowitzsch, Hoicka and van Tulder, 2020), promote local self-organisation (van der Schoor and Scholtens, 2015), or be a tool to stimulate the growth of the local economy (Koirala *et al.*, 2016). Considering the inclusion of local communities in the realisation of tasks and policies created and implemented by public administration, it can be figuratively said that the hitherto model of administration based on the performance of a single actor (the state through administration) is being transformed by the entry of new actor–stakeholders onto the scene. In this context, the term "citizen-centricity" (Pereira *et al.*, 2018, pp. 147–148) fits both the trend of modern administration (not coincidentally, also sometimes referred to with the prefix "smart") and smart cities to characterise the processes.

8.3 Smart Solutions in Japan: Focus on the Key Municipal Sectors

Japan has been consequently introducing smart policy initiatives. Amidst these, projects demonstrating administrative entities' fresh approach to bearing public interest tasks should be recognised and endorsed. The mechanisms, described further, implemented into the practice by public administration bodies could serve as the model for deploying smartness in the context of crucial urban economy sectors. The four selected theme areas are (i) energy, (ii) waste, (iii) water, and (iv) transport. A similar division was made by Kono, Suwa and Ahmad (2016) and Su, Miao and Wang (2022). These sectors were chosen due to their significance in Japan's journey towards creating a sustainable future. The country has articulated a strong motivation to foster smart metropolitan areas that prioritise sustainable development and address pressing environmental concerns (MIAC, METI and MLIT, 2021). Even if legislation does not explicitly mention smart cities, its motives provide a legal and policy framework for their founding (Establishing the Recycling-Based Society Law, 2000). Moreover, ensuring energy independence, in the broader meaning from its production from various sources to possible ways of consumption, is a critical priority for Japan, making energy-related systems a central focus of smart city endeavours (Clarisse, 2014, p. 8). Finally, these four fields are particularly appealing because of their status as essential municipal services, therefore, at the same time being integral and irreplaceable components of urban infrastructure (Clarisse, 2014, p. 8). Given their significance from an administrative management point, among these four fundamental municipality sectors, more smart solutions should be implemented to achieve overall metropolitan advancement in Japan by public bodies. With the presentation of smart applications towards diverse mechanisms by administrative entities, the purpose of this section is also to include how the three main trends in modern administration described earlier are deployed in the smart city concept.

8.3.1 Energy Production and Consumption

Smart city solutions in the energy sector encompass a range of technologies and strategies aimed at every stage of energy lifespan, including production, distribution, consumption, and storage (Sokołowski, 2022). Many initiatives are designed to ensure basic energy needs, especially in the event of unexpected power outages (Strielkowski *et al.*, 2020), on which the emphasis increased after Fukushima event. Thus, the prevailing theme regarding energy-related smart mechanisms evolves around achieving optimal balance while prioritising storage. As an example in Utsunomiya City (n.d.) the local government installs solar panels on public buildings, parks, or schools. Additionally, the town installed advanced metering infrastructure to respond to energy demand and supply as effectively as possible (MIAC, METI and MLIT, 2021). Remedial measures were also taken by the Kashiwanoha. In the area, buildings and infrastructure share power in case of emergencies.

Energy-oriented projects are commonly undertaken with the formula of public–private partnerships due to the costs associated with the network technical development and the expansion into areas not reached by the grid. At the same time, there is a shared interest in business–government partnerships (Pianezzi, Mori and Uddin, 2023, p. 8) on account of the climate transformation strategy and domestic legal acts enforcing sustainable measures (Special Measures Concerning Procurement of Electricity from Renewable Energy Sources by Electricity Utilities Act, 2011; Climate Change Adaptation Act, 2018; Low Carbon City Promotion Act, 2012). On the position of the project contractors, technological corporations or scientific institutions could join to be supported financially by METI or other ministries, such as the Ministry of Land, Infrastructure, Transport and Tourism (MLIT), Ministry of Internal Affairs and Communications) MIAC, or Ministry of Environment (MoE) (see Rondinelli *et al.*, 2008, p. 21). One of the most notable governmental funding was done under the Next-Generation Energy and Social System Demonstration Projects in 2010, where four municipalities, Keihanna, Kitakyushu, Yokohama, and Toyota City, obtained aid (Wongbumru and Dewancker, 2014, p. 73). Smaller administrative units' involvement hinges on issuing formal local energy strategies (e.g. city of Kitakyushu, 2017) or smaller-scale projects for the meagre group of beneficiaries also in cooperation with private parties (Rondinelli *et al.*, 2008, p. 30). The number of these local government initiatives has been increasing significantly each year (DeWit, 2014, p. 10).

8.3.2 Waste Management

The vision of waste management in smart cities relies on a no-waste approach, which could be linked with a circular economy concept (Aceleanu *et al.*, 2019, p. 1). Therein, each produced element has its designated place towards its lifespan. In the motives of sustainability, it is essential to manage their allocation towards efficiency, especially by local government bodies. Failure to carry out precisely the obligations incumbent on local governments risks endangering public health and safety, as was particularly highlighted after the COVID-19 pandemic (Onoda, 2020; Gade, and Aithal, 2021a). With the help of modern "smart" techniques and through the involvement of other stakeholders, the public administration is able to meet one of the basic public governance tasks, that is protecting health and life, even under emergency conditions.

Smart waste management activities are mainly undertaken based on artificial intelligence (AI) and IoT mechanisms or other information and communication technologies (ICT) (Gade and Aithal, 2021b). There are tools for collecting data on the amount of waste in trash cans but also places not intended for litter storage, for example sidewalks or parks. Thus, utilisation services are operated effectively and, above all, allow for reaction to the possible spread of contaminated elements. Another element, this time supporting the energy transition, is the reuse of unused materials as recyclable products or through waste-to-power as a process of energy recovery.

The residential areas able to claim a high level of waste disposal are Yokohama and Osaka, where extensive activities of public bodies, such as technological innovation support, proper sanctions imposition and execution, or climate awareness education, have cut the waste emissions significantly. In the case of the first city, it dropped under 40% between 2003 and 2009 (Hashimoto, 2021), second under 80% compared from 2020 to 1990, which states approximately 1 million tons of waste annually (Fietkiewicz and Stock, 2015, p. 2352).

8.3.3 Water Management

Similar to waste, the use of monitoring tools is applied in water management in smart cities to enhance water efficiency, detect leaks, and optimise water distribution. By leveraging the smart system data processing, mainly with the contribution of AI, it becomes feasible to predict and mitigate natural disaster consequences such as flooding, which frequently impacts Japan (Sánchez-Corcuera *et al.*, 2019). Such management measures, which can be also deemed as remedial, are common in many Japanese cities, for example in Fujieda (Kitamura, 2021) or Masuda (Toyosaki, 2021).

Water, both a source of life and a critical component in numerous industrial processes, is indispensable for sustainable urban development. Water management and its reuse are the foundations of the circular economy. Furthermore, the mechanism of rational administration of natural resources enters as the key element in smart city management theory (Molina-Giménez, 2018, p. 13). This tenet was noted before the aforementioned Next-Generation Energy and Social System Demonstration Projects, the first comprehensive impulse into smart cities development. Hence, the METI, in the early period of 2005–2007, subsidised the project in this area, for example "Project of Energy Return from Flowing Water" in Hachinohe (Fujioka, 2003; Rondinelli *et al.*, 2008, p. 30).

8.3.4 Transportation

Despite the fact that transport is described in the chapter as the last sector, it can be considered the area where the idea of smart administration has the most impact (Docherty, Marsden and Anable, 2018, Table 1). Therein the role of public administration revolves around linking the complex interconnections between investors, authorities entities, passengers, and transport service-providers (Su, Miao and Wang, 2022; Chen, Ardila-Gomez and Frame, 2017; Nguyen *et al.*, 2020; Docherty, Marsden and Anable, 2018). Considering the perplexity of this area, administration bodies should be engaged and coordinate every stage of smart mobility service (Chen, Ardila-Gomez and Frame, 2017), from the design phase, through the concluding arrangements with involved parties, to finally performing transportation duties on a public basis. Thus, the participation of decentralised administration bodies operating in the municipality should prevail over central offices not familiar with the specifics of the relevant territory.

Within this approach, a few interesting solutions could be distinguished in the context of Japanese metropolises. One of the most vibrant and sophisticated agendas is the Kōzōji New Mobility Town Plan. Kōzōji intends to be the first city to offer on-demand autonomous transportation service for private use for free. The service will offer a vehicle for rent, which will be available for any coveted purpose according to the user's request, such as shopping, transporting children to school, or for a leisurely ride (Kōzōji Smart City Promotion Study Committee, 2023). It demonstrates that transportation can be used not only to reach a destination as quickly as possible but can also depict the convenient exploitation of communal means of transport by being customised to individual preferences.

8.4 Conclusion

The development of the smart cities concept as a tool for public administration is inextricably linked to the development of modern methods of public administration and new ways of interacting with administered entities. Undoubtedly, public administrations are taking a more active role in ensuring the highest possible quality of life for residents than just a few decades ago. The example of Japan proves that the implementation of smart city solutions leads to the occurrence of public administrations in this new type of action, leading to the development of solutions that go beyond the standard perception of the relationship between administering entities and administered entities. In fact, the smart city concept is a reality created by all entities, regardless of their place in the administrative structure. This participation of such a wide range of interest groups implements smart city solutions continuously. It leads to an improvement in the effectiveness of public policies by increasing public participation and directing private investment towards objectives in line with the interests of public administrations performing climate and energy transition tasks. Especially within smart cities, sustainability contributes to the greening of municipal sectors, which prioritise meeting citizens' basic needs rather than solely maintaining environmental standards.

Municipal services in smart cities not only meet the basic needs of their citizens but also contribute to improving the quality of life by increasing the convenience, speed, and availability of service delivered. Through advanced technologies and innovative solutions, municipal services, such as smart data management systems, are designed to be more efficient and more responsive to residents' needs. Moreover, the effectiveness of costs is maximised up to the possible limit. By implementing smart solutions in utility sectors, public administrations enhance security by continuously reviewing and adapting functionality mechanisms through the accessed data. This combination of tools spawns foundations for a coherent landscape of a sustainable and self-sufficient economy.

In the case of the involvement ratio of administration bodies in smart city projects, currently, there is still a high degree of centralisation. This is a result of, *inter alia*, access to funds that makes central administrative units crucial with respect to redistribution of financial resources. However, smaller administrative units gradually increase their presence in this field, primarily by active participation in designing the smart city concept in line with trends in administration such as deprivatisation, decentralisation, and deprioritisation.

References

Aceleanu, M.I. *et al.* (2019) "The Management of Municipal Waste through Circular Economy in the Context of Smart Cities Development", *IEEE*, 7, pp. 133602–133614.

Al Nuaimi, E. *et al.* (2015) "Applications of Big Data to Smart Cities", *Journal of Internet Services and Applications*, 6(25), pp. 1–15.

Allam, Z. and Dhunny, Z.A. (2019) "On Big Data, Artificial Intelligence and Smart Cities", *Cities*, 89, pp. 80–91.

Appio, F., Limab, M. and Sotirios, P. (2019) "Understanding Smart Cities: Innovation Ecosystems, Technological Advancements, and Societal Challenges", *Technological Forecasting & Social Change*, 142, pp. 1–28.

Bardhan, P. (2002) "Decentralization of Governance and Development", *Journal of Economic Perspectives*, 16(4), pp. 185–205.

Basu, R. (2004) *Public Administration Concepts and Theories.* Greater Noida: Sterling Publishers Private.

Basu, R. (2019) *Public Administration in the 21st Century. A Global South Perspective.* London: Routledge.

Bland, F.A. (1942) "Public Administration in War-Time", *The Australian Quarterly*, 14(3), pp. 64–76.

Bleja, J. *et al.* (2020) "Smart Cities for Everyone – Age and Gender as Potential Exclusion Factors", *2020 IEEE European Technology and Engineering Management Summit (E-TEMS)*, Dortmund, pp. 1–5.

Callahan, K. (2007) "Citizen Participation: Models and Methods", *International Journal of Public Administration*, 30(11), pp. 1179–1196.

Chen, Y., Ardila-Gomez, A. and Frame, G. (2017) "Achieving Energy Savings by Intelligent Transportation Systems Investments in the Context of Smart Cities", *Transportation Research Part D: Transport and Environment*, 54, pp. 1–40.

Clarisse, P. (2014) "Smart Cities in Japan. An Assessment on the Potential for EU-Japan Cooperation and Business Development". Tokyo: EU-Japan Centre for Industrial Cooperation. Available at: www.eu-japan.eu/sites/default/files/publications/docs/smartcityjapan.pdf

Climate Change Adaptation Act (2018), no. 50.

Cohen, J.M. and Peterson, S.B. (1996) "Methodological Issues in the Analysis of Decentralization", *Development Discussion Paper No. 555*, Cambridge, Harvard Institute for International Development.

Crome, C. *et al.* (2023) "Circular economy Is Key! Designing a Digital Artifact to Foster Smarter Household Biowaste Sorting", *Journal of Cleaner Production*, 423, 138613.

Crosby, N., Kelly, J.M. and Schaefer, P. (2015) "Citizens Panels: A New Approach to Citizen Participation", in Roberts, N.C. (ed.), *The Age of Direct Citizen Participation.* London: Routledge, pp. 266–278.

Dashkevych, O. and Portnov, B.A. (2022) "Criteria for Smart City Identification: A Systematic Literature Review", *Sustainability*, 14(8), p. 4448.

DeWit, A. (2014) “Japan’s Resilient, Decarbonizing and Democratic Smart Communities”, *The Asia-Pacific Journal*, 12(50). https://apjjf.org/volume-and-issue/volume-12/issue-5

Docherty, I., Marsden, G. and Anable, J. (2018) “The Governance of Smart Mobility”, *Transportation Research Part A: Policy and Practice*, 115, pp. 114–125.

Đurman, P. (2020) “Participation in Public Administration Revisited: Delimiting, Categorizing and Evaluating Administrative Participation”, *Hrvatska i komparativna javna uprava: časopis za teoriju i praksu javne uprave* [Croatian and Comparative Public Administration, *Journal for Theory and Practice of Public Administration*, 20(1), pp. 79–120.

Establishing the Recycling-Based Society Law (2000) no. 110. In Japanese: 循環型社会形成推進基本法 [junkan-gata shakai keisei suishin kihon-hō].

Fietkiewicz, K.J. and Stock, W.G. (2015) "How "Smart" Are Japanese Cities? An Empirical Investigation of Infrastructures and Governmental Programs in Tokyo, Yokohama, Osaka, and Kyoto" in *48th Hawaii International Conference on System Sciences*, 5-8 January 2015, Kauai, Hawaii: HICSS, pp. 2345–2354, doi: 10.1109/HICSS.2015.282

Fry, B.R. (1989) *Mastering Public Administration: From Max Weber to Dwight Waldo*. New Jersey: Chatham House Publishers.

Fujioka, Y. (2003) “The Project of ‘Energy Return from Flowing Water’, in Hachinohe”, *Journal of the Gas Turbine Society of Japan*, 31(6), pp. 399–404.

Gade, D.S. and Aithal, P.S. (2021a) “Smart Cities Development During and Post COVID-19 Pandemic – A Predictive Analysis”, *International Journal of Management, Technology, and Social Science*, 6(1), pp. 189–202.

Gade, D.S. and Aithal, P.S. (2021b) “Smart City Waste Management through ICT and IoT Driven Solution”. *International Journal of Applied Engineering and Management Letters*, 5(1), pp. 51–65.

Gavkalova N.I. and Kolupaieva I.V. (2018) "Decentralization of public administration in the process of building a democratic society", *Public Policy and Administration*, 17(2), pp. 217–225, doi:10.13165/VPA-18-17-2-05

General Rules for Incorporated Administrative Agencies Act (1999) no. 103. In Japanese: 独立行政法人通則法 [dokuritsu gyōsei hōjin tsūsoku-hō]

Gjorgievski, V.Z., Cundeva, S. and Georghiou, G.E. (2021) “Social Arrangements, Technical Designs and Impacts of Energy Communities: A Review”, *Renewable Energy*, 169, pp. 1138–1156.

Guseva, M.S. and Dmitrieva, E.O. (2020) “The Competitiveness of Single-Industry Cities in the Digital Transformation of The Economy”, in Ashmarina, S., Mesquita, A. and Vochozka, M. (eds) *Digital Transformation of the Economy: Challenges, Trends and New Opportunities*. Cham: Springer, pp. 216–226.

Hashimoto, T. (2021) “Yokohama’s City to City Collaboration on Circular Economy and Sustainability”. Available at: https://www.thegpsc.org/sites/gpsc/files/yokohama_y_port.pdf (Accessed 17 February 2024).

Heffron, R.J. and Sokołowski, M.M. (2022) “Defining and Conceptualising Energy Policy Failure: The When, Where, Why, and How”, *Energy Policy*, 161, 112745.

Holstein, J.A. and Gubrium, J.F. (1995) “Deprivatization and the Construction of Domestic Life”, *Journal of Marriage and Family*, 57(4), pp. 894–908.

Hori, M. (2003) “Japanese Public Administration and its Adaptation to New Public Management”, *Ritsumeikan Law Review*, 20, pp. 1–15.

Howe, L.E. (2002) “Administrative Law and Governmentality: Politics and Discretion in a Changing State of Sovereignty”, *Administrative Theory & Praxis*, 24(1), pp. 55–80.

Hoye, J.M. (2019) "Natural Justice, Law, and Virtue in Hobbes's Leviathan", *Hobbes Studies*, 32(2), pp. 179–208.

Kakuno, S. (2014) "Overview of NEDO's CCT Development". Available at: www.pl.emb-japan.go.jp/keizai/documents/E2_1%20NEDO%20Kakuno.pdf (Accessed: 17 February 2024).

Khansari, N., Mostashari, A. and Mansouri, M. (2013) "Impacting Sustainable Behaviour and Planning in Smart City", *International Journal of Sustainable Land Use and Urban Planning*, 1(2), pp. 46–61.

Kitamura, S. (2021) "Fujieda City, Shizuoka Prefecture". Available at: www.jasca2021.jp/1st/pdf/discus/Fujieda.pdf (Accessed: 17 February 2024).

Kitchin, R. and Dodge, M. (2019) "The (In)Security of Smart Cities: Vulnerabilities, Risks, Mitigation, and Prevention", *Journal of Urban Technology*, 26(2), pp. 47–65.

Kjaer, P.F. (2018) "From the Private to the Public to the Private? Historicizing the Evolution of Public and Private Authority" *Indiana Journal of Global Legal Studies*, 25(1), pp. 13–36.

Klijn, E. H. and Koppenjan, J. (2016) *Governance Networks in the Public Sector*. London: Routledge.

Koirala, B.P. *et al.* (2016) "Energetic Communities for Community Energy: A Review of Key Issues and Trends Shaping Integrated Community Energy Systems", *Renewable and Sustainable Energy Reviews*, 56, pp. 722–744.

Kono, N., Suwa, A. and Ahmad, S. (2016) "Smart Cities in Japan and Their Application in Developing Countries", in J. Jupesta and T. Wakiyama (eds) *Low Carbon Urban Infrastructure Investment in Asian Cities*. London: Palgrave Macmillan, pp. 95–122.

Kōzōji Smart City Promotion Study Committee (2023) "Kōzōji New Mobility Town Plan", Japan, Kōzōji. Available at: https://uncrd.un.org/sites/uncrd.un.org//files/2023sctw_s3_p2.pdf (Accessed: 17 February 2024).

Kusumasari, B. and Alam, Q. (2012) "Bridging the Gaps: The Role of Local Government Capability and the Management of a Natural Disaster in Bantul, Indonesia", *Natural Hazards*, 60(2), pp. 761–779.

Lodge, M. and Wegrich, K. (2014) "Governance Innovation, Administrative Capacities, and Policy Instruments", in M. Lodge and K. Wegrich (eds) *The Problem-Solving Capacity of the Modern State: Governance Challenges and Administrative Capacities*. Oxford: Oxford University Press, pp. 1–24.

Low Carbon City Promotion Act (2012) no. 84. In Japanese: 都市の低炭素化の促進に関する法律 [toshi no tei tansoka no sokushin ni kansuru hōritsu].

Lowitzsch, J., Hoicka, C.E. and van Tulder, F.J. (2020) "Renewable Energy Communities Under The 2019 European Clean Energy Package – Governance Model for The Energy Clusters of the Future?", *Renewable and Sustainable Energy Reviews*, 122, p. 109489.

Makieła, Z. J. et al. (2022) "Smart City 4.0: Sustainable Urban Development in the Metropolis GZM", Sustainability, 14(6), 3516.

Marrone, M. and Hammerle, M. (2018) "Smart Cities: A Review and Analysis of Stakeholders' Literature", *Business & Information Systems Engineering*, 60, pp. 197–213.

Meijer, A. and Bolívar, M.P.R. (2016) "Governing the Smart City: A Review of the Literature on Smart Urban Governance", *International Review of Administrative Sciences*, 82(2), pp. 392–408.

MIAC, METI and MLITT (2021) "Smart City Guidebook". Available at: www8.cao.go.jp/cstp/society5_0/smartcity/00_scguide_eng_ol.pdf (Accessed: 17 February 2024).

Moe, T.L. and Pathranarakul, P. (2006) "An Integrated Approach to Natural Disaster Management: Public Project Management and Its Critical Success Factors", *Disaster Prevention and Management*, 15(3), pp. 396–413.
Molina-Giménez, A. (2018) "Water Governance in the Smart City", *Urban Growth and the Circular Economy*, 179, pp. 13–22.
Nesti, G. and Graziano, P.R. (2019) "The Democratic Anchorage of Governance Networks in Smart Cities: An Empirical Assessment", *Public Management Review*, 22(5), pp. 1–20.
New Energy and Industrial Technology Development Organization Act (2002) no. 145.
Nguyen, D.D. *et al.* (2020) "Intelligent Total Transportation Management System for Future Smart Cities", *Applied Sciences*, 10(24), p. 8933.
Okubo, H. *et al.* (2022) "Smart Communities in Japan: Requirements and Simulation for Determining Index Values", *Journal of Urban Management*, 11, pp. 500–518.
Onoda, H. (2020) "Smart Approaches to Waste Management for Post-COVID-19 Smart Cities in Japan", *IET Smart Cities*, 2(2), pp. 89–94.
Ospanova, D. *et al.* (2018) "Problems of Administrative Law in the System of Public Administration", *International Journal of Law and Management*, 60(6), pp. 1255–1271.
Paneru, C.P., and Tarigan, A.K.M. (2023) "Reviewing the Impacts of Smart Energy Applications on Energy Behaviours in Norwegian Households", *Renewable and Sustainable Energy Reviews*, 183, p. 113511.
Pereira, G.V. *et al.* (2018) "Smart Governance in the Context of Smart Cities: A Literature Review", *Information Polity*, 23(2), pp. 143–162.
Pianezzi, D., Mori, Y. and Uddin, S. (2023) "Public–Private Partnership in a Smart City: A Curious Case in Japan", *International Review of Administrative Sciences*, 89(3), pp. 1–16.
Piecha, J. (2022) "Social Control of Public Administration: Twilight or Renaissance? Observations in the Context of the Administration Performed Through the Algorithms", *Studia Iuridica*, 89, pp. 269–286.
Pound, R. (1941) *Administrative Law: Its Growth, Procedure, and Significance*. Pittsburgh, PA: University of Pittsburgh Press.
Purnendra, J. (2002) "The Catch-up State: E-government in Japan", *Japanese Studies*, 22(3), pp. 237–255.
Rondinelli, D.A. (1990) "Decentralizing Urban Development Programs: A Framework for Analyzing Policy", Agency for International Development. Office of Housing and Urban Programs, Research Triangle Institute.
Rondinelli, D.A. *et al.* (2008), *Public Administration and Democratic Governance: Governments Serving Citizens*. New York: United Nations Publication.
Sánchez-Corcuera, R. *et al.* (2019) "Smart Cities Survey: Technologies, Application Domains and Challenges for the Cities of the Future", *International Journal of Distributed Sensor Networks*, 15(6), pp. 1–36.
Schneider, A. (2003) "Decentralization: Conceptualization and Measurement", *Studies in Comparative International Development*, 38, pp. 32–56.
Shimpo, F. and Mikusek, P. (2023) "Personal Data Protection Issues in the Context of Smart Cities Solutions" in Sokołowski, M.M. and Visvizi, A. (eds), *Routledge Handbook of Energy Communities and Smart Cities*. London: Routledge, pp. 205–219.
Shiroyama, H. and Kajiki, S. (2016) "Case Study of Eco-town Project in Kitakyushu: Tension among Incumbents and the Transition from Industrial City to Green City", in Loorbach, D. *et al.* (eds), *Governance of Urban Sustainability Transitions. European and Asian Experiences*. Tokyo: Springer, pp. 113–132.

Sokołowski, M.M. (2021) "Models of Energy Communities in Japan (Enekomi): Regulatory Solutions from the European Union (Rescoms and Citencoms)", *European Energy and Environmental Law Review*, 30(4), pp. 149–159.

Sokołowski, M.M. (2022) "Artificial Intelligence and Climate-Energy Policies of the EU and Japan", in M. Bielicki, (ed.) *Regulating Artificial Intelligence in Industry*. London: Routledge, pp. 138–155.

Sokołowski, M.M. and Kurokawa, S. (2022) "Energy Justice in Japan's Energy Transition: Pillars of Just 2050 Carbon Neutrality", *Journal of World Energy Law and Business*, 15, pp. 183–192.

Sokołowski, M.M. and Visvizi, A. (eds) (2023) *Routledge Handbook of Energy Communities and Smart Cities*. London: Routledge.

Soni, S. and Bhushan, B. (2019) "A Comprehensive Survey on Blockchain: Working, Security Analysis, Privacy Threats and Potential Applications" in *2nd International Conference on Intelligent Computing, Instrumentation and Control Technologies (ICICICT)*, 1, pp. 922–926.

Špaček, D., Navrátil, M. and Špalková, D. (2023) "New Development: Covid 19 and Changes in Public Administration – What do we Know to Date?", *Public Money & Management*, 43(8), pp. 862–866.

Special Measures Concerning Procurement of Electricity from Renewable Energy Sources by Electricity Utilities Act (2011), no. 108.

Sta, H.B. (2017) "Quality and the Efficiency Of Data in 'Smart-Cities'", *Future Generation Computer Systems*, 74, pp. 409–416.

Steffan, W. *et al.* (2015) "Planetary boundaries: Guiding human development on a changing planet", *Science*, 347(6223), p. 1259855.

Strielkowski, W. *et al.* (2020) "Economic Efficiency and Energy Security of Smart Cities", *Economic Research-Ekonomska Istraživanja*, 33(1), pp. 788–803.

Su, Y., Miao, Z. and Wang, C. (2022) "The Experience and Enlightenment of Asian Smart City Development – A Comparative Study of China and Japan", *Sustainability*, 14(6), p. 3543.

Sundar, N. (2001) "Is Devolution Democratization?" *World Development*, 29(12), pp. 2007–2023.

Torgerson, D. (1990) "Obsolescent Leviathan: Problems of Order in Administrative Thought", *Managing Leviathan: Environmental Politics and the Administrative State*, pp. 17–34. https://doi.org/10.3138/9781442602281-006

Toyosaki, Y. (2021) "Water Level Monitoring Project Outline". Available at: www.jasca2021.jp/1st/pdf/WS1/Masuda_City.pdf (Accessed: 17 February 2024).

Utsunomiya City (n.d.) "Utsunomiya: Your Convention City". Available at: https://utsunomiya-convention.jp/en./ (Accessed: 17 February 2024).

Valdez, A.M. *et al.* (2021) "Learning Lessons for Avoiding the Inadvertent Exclusion of Communities from Smart City Projects", in A. Aurigi and N. Odendaal (eds) *Shaping Smart for Better Cities. Rethinking and Shaping Relationships Between Urban Space and Digital Technologies*. London: Academic Press, pp. 221–237.

van der Schoor, T. and Scholtens, B. (2015) "Power to the People: Local Community Initiatives and the Transition to Sustainable Energy", *Renewable and Sustainable Energy Reviews*, 43, pp. 666–675.

Vocca, J. (2020) *Solving Urban Infrastructure Problems Using Smart City Technologies: Handbook on Planning, Design, Development, and Regulation*. San Diego: Elsevier.

Wang, Q. and Chen, X. (2013) “Rethinking and Reshaping the Climate Policy: Literature Review and Proposed Guidelines”, *Renewable and Sustainable Energy Reviews*, 21, pp. 469–477.

Wong, M.S., Hideki, N. and George, P. (2011) “The Use of Importance-Performance Analysis (IPA) in Evaluating Japan’s E-Government Services”, *Journal of Theoretical and Applied Electronic Commerce Research*, 6(2), pp. 17–30.

Wongbumru, T. and Dewancker, B. (2014) “Smart Communities for Future Development: Lessons from Japan”, *International Journal of Building*, 3, pp. 69–75.

World Economic Forum (2024) *The Global Risks Report 2024. 19th Edition. Insight Report.* Cologny – Geneva: World Economic Forum Publication.

Yamaguchi, Y. (2019) “A Practice-Theory-Based Analysis of Historical Changes in Household Practices and Energy Demand: A Case Study from Japan”, *Technological Forecasting and Social Change*, 145, pp. 207–218.

Yarime, M. (2020) "Facilitating Innovation for Smart Cities: The Role of Public Policies in the Case of Japan” in Y. Joo *et al.* (eds) *Smart Cities in Asia.* Cheltenham: Edward Elgar Publishing, pp. 93–106, doi: 10.4337/9781788972888

9 Smart Cities as a Sustainable Development Tool in Spatial Planning Acts of the European Union and Japan

Comparative Analysis

Piotr Zieliński

9.1 Introduction

Despite cultural differences and physical distance, Japan and the European Union (EU) have strong economic and geopolitical ties, as well as a similar vision for the future. Both sides see the need of building competitiveness through innovation, providing development subsidies, promoting modern technologies, and protecting intellectual property, as evidenced by their ongoing high ranking in the Global Innovation Index. At the same time, both are particularly committed to sustainable development, which is reflected in the internationally outstanding assumptions derived from the EU Green Deal, Clean Energy Transition, and Japan's Eco-Point Programme striving to create an ecologically neutral, energy-saving, and thus efficient and advantageous economy. These aspirations align with the concept of a "Smart City", which combines the postulate of improving quality of life with digital transformation.

From 1990s to 2000s, the idea of digital cities, which were the early stages of smart cities, gradually influenced urban development in Europe and Japan (Anthopoulos, 2017). Although the concept of smart city is relatively new, having only been around for two decades, it has quickly gained popularity among researchers in various fields of social activity, and in 2016, it was one of the vital concepts presented at the United Nations Conference on Housing and Sustainable Urban Development, transferring its assumptions to public policy and becoming a significant tool for sustainable spatial planning.

9.2 The Meaning of "Smart City" in the Public Policy of Japan and the European Union

According to the European Commission (2024), a smart city is a place where traditional connections and cooperation are more effective, thanks to the use of digital solutions for the benefit of both residents and businesses. The means for introducing this concept are the use of modern technologies, better resource allocation,

DOI: 10.4324/9781003471417-9

and emission reduction through, among other things, intelligent urban public transportation, modernised water and waste management devices, more efficient lighting and heating systems. Such activities include more interactive and responsive city administration, safer public spaces, and tackling the negative effects and social challenges associated with an ageing society.

The Japanese Cabinet Office presents a tighter, but praxeologically similar, concept, indicating that a smart city can take various forms depending on the region or nature of the problem at hand, but its general axiological foundation is the assumption that the central nexus of the city is its inhabitants, and that improving their well-being is the primary goal of each urban policy. Japanese policymakers define the means to achieve smart cities, ranging from the deployment of new urban technologies to the use of various data to educational, social, and planning activities, the appropriate selection of which creates new added value, transforming the city's development into an innovative and, at the same time, sustainable one. They demonstrate that it is critical to consistently identify dynamic challenges that alter with technological advancement and then implement relevant solutions (Prime Minister's Office of Japan, 2021).

The intergovernmental EU–Japan Centre for Industrial Cooperation's Report (2014), which examines the potential of smart cities for EU–Japan cooperation and commercial development, provides common ground for both views. Although the report is based on definitions presented by the Japan Smart Community Alliance and the European Parliament prior to 2014, due to their scope and terminological convergence with the concepts currently presented in both institutions' documents, the conclusions of this comparison are still largely valid. The study explicitly states that, due to the broadness of the issue, designating a smart city may be difficult at the regional, let alone international, level. For example, Fujitsu defines it as a concept based on the use of technology for the effective development of infrastructure centred on the comfort of city use, whereas Yokohama authorities define it as a model aimed primarily at dealing with the challenges of climate change and an ageing society. Furthermore, the Fukushima strategy shifted the path to developing a consistent idea of this term in Japan, exposing energy security concerns. This chapter examines the differences between the Japanese and European approaches to the concept of a smart city, concluding that the European model is broad-based and encompasses many fields, referring to various areas of societal functioning, whereas the Japanese model is slightly narrower, more specific, and focuses primarily on energy, infrastructure, lifestyle, and information and communication technology (ICT). At the same time, the chapter highlights important commonalities between the European Union's and Japan's approaches to shaping future cities through innovation. In both cases, ICT is an important tool for implementing new solutions in the city, from an economic, social, or environmental standpoint. New developments in smart cities make energy consumption increasingly reliant on local and renewable sources while also becoming more effective and measurable. Among others, electric public transport, self-sufficient infrastructure elements, and data control centres transform the urban structure of cities by integrating their individual elements, while the use of modern techniques such as CCTV, rapid alarm systems, and solar lighting improves the safety of many urban areas.

The aforementioned EU and Japanese authorities, while situating and defining the concept of a smart city in the structure of social life, recognise that technical advancement is not a value in and of itself, because its primary purpose is to improve people's quality of life. This is noteworthy because many trends in the debate on the concept of a smart city have focused primarily on technological conditions (Angelidou, 2014), whereas the uniqueness of this concept lies in the ability to build development through a combination of material and social capital (Correia and Wünstel, 2011).

However, it is relevant to note that, while the reference point is the improvement of human life quality, this relationship does not imply that the beneficiary of each upgrade automatically becomes a city user belonging to future generations or, for a variety of reasons, is already excluded from selected social goods. In this context, this concept should also be applied to the unresolved problems of ordinary life for all, as well as the continuous civilisational challenges confronting the global community, including Japan and the EU countries. In this regard, the multidimensional sustainable development agenda and the United Nations' Sustainable Development Goals (SDGs) established in 2015 could serve as a point of reference. According to Ahvenniemi *et al.* (2017), the core objective of a smart city is to increase sustainability through the use of technology, and they even suggest replacing it with the term "Smart Sustainable City", which appears in UN publications already.

While the actual contribution of a smart city to meeting civilisation challenges is still being investigated, there is no doubt that this concept provides an opportunity not only to integrate technological, social, and economic conditions but also to direct them towards existing challenges (Schraven, Joss, and de Jong, 2021; Sharifi, 2019). However, authors such as Yigitcanlar and Kamruzzaman (2018) highlight the risks associated with implementing this idea, such as the danger of increased CO_2 emissions. It may be contended that while well-managed smart cities can be an engine for sustained development, both concepts cannot be considered equally. Individual solutions, regardless of country, should be evaluated comprehensively, taking into account both good and negative outcomes.

9.3 The Importance of Smart Sustainable Solutions in Spatial Planning

Spatial planning is a critical area for merging the ideas of smart cities with sustainable development for a variety of reasons. To begin, the transformation of a city cannot be adequately explained without considering its physical and spatial characteristics, as well as the restrictions and factors that accompany them. The territorial allocation of urban activities, forms, and functions, as well as their mutual connections, are crucial for maximising city resources and achieving efficient and sustainable growth (Bourdic, Salat, and Nowacki, 2012). Utilisation of modern technologies may be beneficial or even required for such actions. In this regard, for example, ICT approaches provide modern urban design and operational solutions, such as drone-based sensing, deep learning, blockchain, or AI-based systems for monitoring city functions (Sharifi *et al.*, 2024; Parmentola *et al.*, 2022). 3D urban models present potential for comprehending spatial scenarios such as pollution, land use, urban forms, and even socio-economic phenomena that occur in geography (Santos *et al.*, 2021). At the same time, authors such as Yigitcanlar, Foth, and Kamruzzaman (2019) note that an overly technocratic approach to innovation in spatial planning may deprive it of the ability to respond to contemporary problems.

Second, cities act as a unique driver for change in the modern world, not only functionally, but also spatially, as seen by ongoing urbanisation. According to 2019 forecasts, built-up areas in Italy, Germany, and Poland will grow by 144, 128, and 121 thousand hectares, respectively, by 2030 (European Commission, 2019). This is significant in terms of organising urban growth tendencies, which frequently extend to ecologically sensitive coastal areas, deltas, river basins, and environmentally fragile regions (Balogun *et al.*, 2020). Appropriately deployed technology may support measurement and optimisation of city tissue boost while also mitigating negative effects on vulnerable territories.

At the same time, Eurostat estimates that the European Union's urban population is expected to grow by 7% between 2020 and 2050. In Japan, although the total number of urban people will fall by 11% during the same period, this is due to a general population loss, as the drop in rural areas will be as much as 45% higher. This means that, as in the European Union, in Japan there will be a definite outflow of people from rural areas to cities, accompanied by a demographic crisis. According to predictions, the EU's population will fall by approximately 2% between 2020 and 2050. Although the natural decline is substantially higher, reaching 8% over time, it is counterbalanced by rapid migration. For Japan, whose immigration systems are not balanced due to their small scale, the total population drop might be as high as 19%. According to the same analyses, the European Union and Japan will have 30% and 38% of the population over 65 years in 2050, respectively. According to the EU–Japan Demographic Joint Study, policymakers in the European Union and Japan recognise smart cities' unique supporting role in mitigating the negative effects of these processes, including ICT support in adapting city infrastructure to the needs of older people and increasing the transport accessibility of medical and care centres. One of the key objectives is to make post-industrial neighbourhoods more appealing to persons with special needs, as well as to revitalise and transform collective housing facilities, which will become more important and desirable (European Commission, 2021).

Third, the ongoing green shift focuses on cities, and consequently urban infrastructure and buildings. According to estimates, metropolitan areas account for 70% of worldwide CO_2 emissions, with transport and buildings contributing the most (IPCC, 2022). In May 2021, the European Union and Japan joined the Green Alliance to accelerate their respective transitions to a climate-neutral, circular, and resource-efficient economy in the next decades. In this context, initiatives to reduce greenhouse gas emissions and anthropogenic pollution are directly linked to spatial policy. Many European cities, including Stockholm, Copenhagen, and Amsterdam, are already constructing public spaces based on "green urbanism", in which the city's green tissue is an important component in creating its spatial identity while also fitting into the country's climate policy. Such tendencies, however on a smaller scale, can also be seen in Japanese cities, such as innovative design solutions in Tokyo (Asgarzadeh *et al.*, 2010). This concept is founded not only on pragmatically recognised emission reduction but also on the creation of green social spaces that consider architectural anthropology and psychophysical circumstances of urban living. In this regard, many technologies that enable social engagement in overcoming climate change while also tokenising carbon credits may be considered promising tools for developing sustainable development (Mora *et al.*, 2021). Furthermore, well-organised transit can increase accessibility for the elderly, children, and women while simultaneously reducing the number of vehicles on the road, which has a significant impact on CO_2 emissions (Soe, 2020).

Given the circumstances described previously, it is fair to say that the city possesses a unique power to lead important changes. In this regard, a multi-level systemic approach to smart cities that takes into account the city's spatial nature is essential. Furthermore, cities must be viewed and treated as urban ecosystems made up of physical, ecological, and social components (Nilon, Berkowitz, and Hollweg, 1999). This assumption should be particularly important for policymakers and lawmakers who determine the desired directions of the state's influence, local officials who have planning authority, and investors who ultimately conduct investment processes in the city. Ahvenniemi *et al.* (2017) rightly points out that "the purpose of the smart city assessment framework is to give guidance for decision-making, enable target setting for cities as well as whether the development is proceeding towards the wanted direction".

9.4 Smartness in the Spatial Planning System

The dynamically changing reality, as well as the variability of factors affecting the causative capacities of individual countries, mean that policymakers must not only constantly monitor the ongoing changes but also give them their proper place in public life. It is necessary to identify individual variables and properly embed them in the administrative framework that organises the social structure. This responsibility is multifaceted, encompassing not only functionality but also time. These frameworks can take a variety of forms and regulatory powers depending on their importance, opportunities, and potential threats. Starting with financial support for specific initiatives, through the development of legal regulations that define their formal framework, to the formulation of long-term state policies. However, it should be noted that "planning" is, by definition, about the future, and urban planning should be proactive rather than reactive (UN Habitat, 2022). In the realm of spatial planning, there are two distinct manifestations of state activity. The first are legal acts that, when they enter into force, establish a specific procedural environment for the implementation of new solutions while also indicating the current approach and viewpoint of public authorities. The second are official strategies, which, while not equivalent to applicable law, demonstrate the intended direction of public activities and generally favoured state policy. While the distinction between these two types of public authority products appears obvious, it is worth noting that the content they produce differs fundamentally.

Legal acts are generally directed at public administration entities and public or private participants in specific investment processes. In democratic states governed by law, such as Japan and the EU countries, each activity of public authorities must be based on the law, which defines the scope of administration activity and outlines its tasks. An instance is local governments' zoning of public space, which applies general principles established by the central legislature to specific local conditions, such as the resource-effective management of the city's infrastructure. Such assumptions have the potential to support smart solutions, such as those in water management and energy use, as well as prevent the negative effects of such innovations, for example, by predicting their exogenous impact beforehand.

The structure of norms addressing non-public users of space differs, as they take the form of bans or limitations on the investor's permissible freedom in activities and are generally abstract in nature. A possible obligation of an investor or property owner to raise a specific device or use a specific technology, for example, to protect against excessive noise or heat loss, is typically secondary to the construction of the main project and serves to protect other assets or spaces threatened by the addressee's activity or omission. Similarly, orders for conducting specific formal activities, such as an impact assessment, are the result of other activities initiated by an investor during the construction process. Legal demands in this field are primarily concerned with the protection of specific goods, such as safety, environmental protection, and spatial order.

As a result, there are no directive norms in the law that co-define the investment process for implementing modern technologies in an authoritative, comprehensive, and coherent manner within the context of the smart city idea. While an examination of legal acts adopted by many countries reveals that many of them refer to the essence of this concept, the majority of them are of a sub-legislative and dispersed nature, or refer to specific solutions and projects treated as intelligent (Tikhaleva, 2023). A review of key spatial planning regulations in Japan and those issued by EU legislators reveals that the term smart city does not appear in these acts. Furthermore, the multidimensional nature of smartness makes it difficult to develop a coherent normative structure for implementing this concept in the traditionally established spatial planning or construction processes. In this sense, the launching of innovative smart solutions in spatial planning arises within the scope of the investor's decision-making freedom during the overall construction process, correlated or stimulated by appropriately targeted activities of public authorities, particularly local government planning actions, and functional or territorial facilitations adopted by national legislators. An example of a functional mechanism is the EU Directive 2009/33/EC on the promotion of low-emission vehicles or Directive 2018/844 on the energy efficiency of buildings, while an instance of a territorial law could be the Japanese National Strategic Special Zones, adopted in 2013 and significantly amended in 2020. Although these acts do not directly address spatial planning, they have a considerable impact on the decisions and solutions taken by entities shaping the city's urban tissue, particularly innovative, effective, and forward-thinking ones. Similar solutions could be found in laws on environmental and climate protection, waste management, the planning of coastal and border areas, protected areas, and others. Innovations can lead to the creation of new public value, generated by the government through legal regulations and other activities (Anthopoulos, 2017). However, through legislative action, public authorities can not only facilitate or stimulate the implementation of smart solutions in construction activities but also constrain, limit, and formalise this process, such as by monitoring or evaluating the expected consequences of their introduction. Furthermore, if the legislature ignores particular issues or regulates them insufficiently or inappropriately, certain public goods may suffer more or less permanent damage. What is crucial is that the proper and responsible gradation of values in the legal system, as well as the harmonisation of individual legal provisions, can ensure that the implementation of innovative solutions in the city's tissue not only respects the assets that need to be protected but also induces sustainable development and progress.

The essence and character of the strategies differ, especially given their multiplicity and diversity. National and international plans are the most wide and generic in nature since they encompass the entire extent of the state apparatus's activities. These documents support the system of geographical links between distinct regions and settlement units, allowing them to fully realise their economic potential through existing and planned economic policies and big infrastructure. They describe the country's development trends, challenges, and opportunities, highlight the primary land uses, and propose policy options and measures, to promote balanced urban and rural development in a coherent, sustainable, and inclusive manner (UN-Habitat, 2022). Although these documents do not directly generate the procedural conditions for creating the investment process or implementing smart planning solutions, they serve as axiological guides for public administration operations and reflect the political atmosphere for enforcing potential adjustments.

9.5 Smart Sustainable Solutions within the Laws that Determine the Spatial Planning Process

At the outset, it should be highlighted that an analysis of spatial planning systems in Japan and the European Union cannot be comprehensive due to the different legal nature of both legislators and the scope of regulation of the acts they create. Japan, with its entrenched legal tradition and solid administrative order, has developed a unified spatial planning system. The law covers national, prefectural, and municipal planning. The acts that define the spatial planning process include the National Spatial Planning Act (1950), the Landscape Act, and the National Land Use Planning Act (1974). The City Planning Act (1968) is particularly important in urban planning since it provides "City Planning Areas", which are zones subject to certain arrangement procedures, transformations, and limits.

At the same time, it uses the terms "urbanisation-promoting area" and "urbanisation-restricted area". The first refers to territories that are already part of the urban tissue and should eventually be fully urbanised, whilst the second refers to areas where urbanisation activities should be strictly controlled. Depending on the intended use, particular investment criteria are supplied for a certain area, as well as infrastructure, communication, social, and environmental solutions, as detailed in the local government's City Planning Master Plan or Municipal Master Plan. The technical dimensions and criteria that individual projects must meet are governed by specific building construction and technology rules, such as the Building Standard Act (1950) and the Urban Renewal Act (1969).

The European Union, unlike Japan, is not a state and does not have a single unified spatial planning system or regulation, which frequently leads to allegations that the European Union lacks competence in spatial planning. This approach ignores the significant influence of EU sectoral policies on spatial development patterns, which might be interpreted as constituting a *de facto* EU spatial planning policy (Dühr, Colomb, and Nadin, 2010). Issues such as territorial cohesion, environment, transport, and energy are shared competences, and legal acts enacted in these areas determine individual Member States' planned solutions. To determine the relevance of EU regulations, the results of the ESPON COMSPASS project (2018) may be useful, in which delegates from various Member States identified the areas of EU policies and legislation that have the biggest impact on their nations' spatial policy. These were considered to comprise rules in the sectors of environment, energy, cohesion, rural development, territorial cooperation, and urban and transportation policy. At the same time, the European Committee of the Regions Commission for Territorial Cohesion Policy and EU Budget (2018) conducted a query on EU spatial planning regulations, indicating that legal acts pertaining to the environment, energy, competition, maritime spatial planning, as well as intergovernmental cooperation strategies in the fields of urban and spatial policy, are critical in light of the United Nations Sustainable Development Goals and New Urban Agenda. Undoubtedly, some of these areas significantly overlap with matters related to the implementation of smart city solutions. Correia, Marques, and Teixeira (2022) conducted a cluster analysis empirically showing the connection between smart city initiatives in the European Union and individual functional areas, which may be useful in finding a relevant comparison for the aforementioned matters. According to the analysis results, the following categories were most frequently related with these initiatives: (1) community; (2) participation and inclusion; (3) energy and lighting; (4) mobility and transportation; (5) digitisation and interoperability; (6) urban planning, and (7) environment and air quality. It can therefore be concluded that spatial planning in the city is one of the leading areas correlated with the idea of a smart city, and what is more, it is among the issues covered by the legal competences of the European Union that are important for its territorial and spatial policy. In this context, identifying the role of specific EU spatial planning regulations associated with the concept of a smart city while also allowing for the implementation of sustainable development requires an examination of legal acts pertaining to environmental protection, energy, and cohesion policy. While the last of the aforementioned categories is primarily based on executive and support tools such as the ESI Funds, Community-Led Local Development, and Integrated Territorial Investment Policy, EU legislation governs environmental protection, climate, and energy. The leading legal acts in this area include, among others, directive on the assessment of the effects of certain plans and programs on the environment (SEA Directive), directive on the assessment of the effects of certain public and private projects on the environment (EIA Directive), directive establishing a framework for community action in the field of water policy (Water Framework Directive), directive on the assessment and management of flood risks, directive on waste and repealing certain directives, directive on the landfill of waste, directive on the promotion of the use of energy from renewable sources (Renewable Energy Directive), directive on energy efficiency (Energy Efficiency Directive), and, to some indirect extent, directive establishing an infrastructure for spatial information in the European community (Council Directive 1999/31/EC, 1999). The stipulations of some of these acts are particularly noteworthy and warrant further investigation.

Thus, the SEA Directive demands an environmental impact assessment for plans and projects that may have a major influence on the general environment. These include planning projects that may encompass a wide range of investment sectors, such as energy, industry, transportation, waste management, water management, telecommunications, tourism, town-country planning, or land use, as well as impact biodiversity, population, human health, fauna, flora, soil, water, air, climatic factors, material assets, and cultural heritage, including architectural and archaeological heritage. This law recommends that planners incorporate environmental factors early in the planning and idea creation process, emphasising the need for public consultation as well. In the context of smart cities, such an integrated approach can help ensure that smart solutions not only strengthen efficiency and technological innovation but also improve residents' quality of life and protect the environment, as outlined in SDGs 11 and 17 on sustainable cities and communities. Another important act is the EIA Directive, which requires an environmental impact assessment for specific investment projects that may have a significant environmental impact. The evaluation needed of investors must not only take into consideration ordinary local conditions but also explain whether there is a possibility of incorporating other, more environmentally conscious solutions, which is a real motivator to seek out new intelligent solutions. It is worth emphasising that for complicated projects, impact assessment can be functionally integrated into the strategic urban planning process, helping to achieve SDG 11 by fostering inclusive, safe, resilient, and sustainable cities.

Similar standards are established in Japan under the Environmental Impact Assessment Act (1997), which governs the environmental effect assessment of public and private projects and programmes. Similar to the EU acts, it mandates early effect forecasting, public participation in the evaluation process, and project differentiation based on scale and nature of impact. Although the scope of regulation varies amongst the legislation previously mentioned, all outline the requirements for environmental evaluation. In this context, the EU Directive requires a more comprehensive review, considering potential climate change, air and water quality issues, population situations, local identity, and archaeological assets. It is worth noting that the obligation to include such factors in spatial planning could not be found in the Japanese City Planning Act (1968), relating directly to spatial planning, as well. This is significant since the city's massive increase in pollution requires finding efficient ways to monitor air quality (Binbusayyis *et al.*, 2024). While indicated issues may seem secondary from the perspective of urban planning, their omission does not support the search for integrated technologies that would take into account all environmental factors. New solutions should be oriented towards a cost-efficient and resource-efficient approach, but they should examine not only the impact on the nature–environment but all environmental factors affecting the quality of life (Ahvenniemi *et al.*, 2017). Moreover, it is worth pointing out that although both EU and Japanese EIA acts are axiologically based on the idea of environmental protection, which is crucial for SDGs 11, 13, and 15, neither of them mentions sustainable development as a comprehensive evaluation criterion broader than the natural environment.

Another EU act important from the perspective of spatial planning is the Water Framework Directive, which aims to protect and improve the condition of water and at the same time favours the construction of integrated water management. The Directive imposes an obligation to establish monitoring programmes, points out the role of appropriate data analysis, and promotes water-efficient technologies and saving techniques. At the same time, it encourages the inclusion of public consultations on water management. It should be assessed that the norms and demands of the Directive favour the construction of modern water infrastructure in cities, the installation of intelligent solutions, and their inclusion in the tasks of designing water networks, locating hydrotechnical devices, and planning coastal areas. Such assumptions are significantly in line with the smart city concept and, at the same time, contribute to improving the protection of aquatic ecosystems in the course of planning processes, which meets the challenges set out in SDGs 6 and 11. In Japan, water management issues are extensively regulated in many legal acts, including the Water Pollution Prevention Act (1970), the Act on Services Related to Waterways (1950), and the River Act (1964). Although there are no clear references to sustainable development in these documents, the solutions adopted by Japanese legislators assume a holistic approach to water management, just like in the European Union, building an integrated water infrastructure, and taking into account population and environmental conditions, which certainly facilitates the implementation of effective and innovative solutions, also in the planning sphere. The City Planning Act (1968) also assumes that aspects related to optimised management of water and reservoirs should be included in the planning process.

An important legal act in the European Union relating to energy policy and influencing the planning process is the Energy Efficiency Directive, which specifies the requirements for energy efficiency. The Directive assumes that reducing energy consumption in the areas of transport, public buildings, spatial planning, water management, public lighting, and infrastructure planning, among others, is an essential element of building energy efficiency. According to the legislator's assumptions, improving energy performance has the potential to foster urban regeneration, including the improvement of buildings and changes in mobility and accessibility patterns. It includes promoting more efficient vehicles, a modal shift to cycling, walking and, because of that, urban plans supporting collective transport. The Renewable Energy Directive is another act that sets strict guidelines for factoring in the potential of renewable energy sources and guaranteeing favourable conditions for their development in spatial planning. This act states that local and regional governments should support self-consumption communities, evaluate the possibility of utilising renewable sources, and include the use of unavoidable waste heat and cold in the planning process, including early spatial planning, designing, building, and renovating urban infrastructure, industrial, commercial, or residential areas, and energy infrastructure. According to the Directive, an evaluation of renewable energy sources must consider the feasibility of small-scale renewable energy production as well as a spatial analysis of areas that are suitable for low-ecological-risk deployment. In Japan, legislators have introduced solutions to optimise energy consumption through laws such as the Act on Special Measures Concerning the Promotion of New Energy Usage (1997), the Act on the Rational Use of Energy (1979), and the Act on Special Measures Concerning the Procurement of Electricity from Renewable Energy Sources by Electricity Utilities (2011). Although the acts establish specific solutions targeted at combating climate change and protecting the local environment, they also contribute to the formation of a recycling society by increasing energy consumption efficiency. The laws emphasise solar and wind power, the necessity for effective use of fossil fuels, and the importance of implementing appropriate energy solutions to revitalise local areas. Nonetheless, these acts make no direct references to spatial planning and make no mention of optimising energy use in the design processes defined in the City Planning Act (1968), which likewise omit elements of energy efficiency. Although these Japanese acts imply the adoption of solutions that can aid smart and sustainable planning solutions, such a link has yet to be fully articulated and described in urban creation terms.

Such advancements have relevance in spatial planning given that they might improve energy effectiveness and the safety of space, boosting its value and investment potential, for instance, through lighting and monitoring systems in places beyond regular urban tissue. It can also help enhance the accessibility of particular areas for diverse social groups who have previously been excluded due to mobility issues (Sharifi *et al.*, 2024). Finally, pro-smart, properly formed legislative solutions might facilitate the rise of entire energy communities, which, by supporting the sustainable redevelopment of industrial zones, contribute to the process of urban improvement (Sokołowski and Visvizi, 2023). In contrast, innovative solutions may have adverse effects, like the impact of artificial light on animals and their habitats, causing birds to migrate prematurely or late in relation to their biological calendar (Foth and Caldwell, 2018). However, there is no doubt that the right selection of such options can result in an integration of inventive smartness and the fulfilment of the environmental, social, and energy goals specified in SDGs 7, 11, 12, and 13. Furthermore, it may contribute to achieving "energy justice". Prioritising this concept will allow for a just energy transition that takes into account human rights, ensures the real application of the SDGs and, above all, leaves no one behind (Sokołowski, 2022; Sokołowski and Kurokawa, 2022). Moreover, legal regulations in the field of energy policy under the smart city concept can significantly impact the situation of city residents and contribute to poverty reduction (Tito *et al.*, 2023).

9.6 Smart Sustainable Cities in Spatial Planning Strategies

Procedures and binding norms set by lawmakers are distinct from the desired direction of state policy and macro-scale issues seen by policymakers and articulated in official government strategies. The National Spatial Strategy (2023) is Japan's main spatial policy document, and it is typically modified every eight years, with the most recent revision being in July 2023. Because of the European system's complexity, it is difficult to identify a single act that would serve as a comprehensive counterpart to such a policy at the EU level. However, the policy paper on spatial planning most frequently cited by experts is Territorial Agenda EU, which outlines mainstream spatial vision and, in theory, is utilised afterwards in Member States' strategic documents (Degórska, 2021).

In the context of implementing smart solutions, documents adopted as part of intergovernmental cooperation between EU Member States, such as the Amsterdam Pact (2016) and the Leipzig Charter (2020), as well as documents prepared under the EU's Urban Agenda, such as the Digital Transition Action Plan, may be instructive. Meanwhile, it should be noted that they are primarily auxiliary and contribute to the general growth of cities. Thus, the EU Territorial Agenda 2030, approved in 2020, recognises the importance of territorial development and spatial integration in boosting innovation and competitiveness, as well as the climate, urbanisation, and socio-economic concerns that Europe faces. Documents recognise that digitalisation may assist in solving spatial problems, contribute to environmental protection, and reduce inequalities. In this context, it refers to smart transportation, e-services, e-governance, and an all-encompassing spatial system. The Leipzig Charter addresses the subject more directly, stating that digitalisation is the primary driving force of transformation and an inter-sectoral trend influencing all aspects of the city's development, including environmentally aware construction. Although the document stresses the necessity of investing in innovative and efficient technologies to improve use of space and accessibility criteria, it outlines possible connections between these areas in a very broad way. The Amsterdam Pact underlines the importance of cross-sectoral collaboration and impact assessments in city development. It emphasises the benefits of the digital transition related to spatial planning, but mainly in terms of strengthening public services for citizens, particularly the disabled and elderly, and fostering business opportunities in urban areas. The Digital Transition Action Plan places the greatest emphasis on smart and digital solutions, indicating that a decision-making process based on properly used data, which includes city officials' expertise, citizen opinions, and research-based knowledge, can transform urban planning processes in order to co-create the future of European cities in the smartest possible way. It indicates the possible benefits of Planned Land Use Data for urban zoning, planning standards, and collaboration among stakeholders. Unfortunately, the recommendations for digitalisation in spatial planning are missing clear and direct linkages to dealing with particular challenges posed by sustainable development.

The Japanese National Spatial Strategy (2023) demonstrates a more comprehensive approach. According to this document, spatial planning should enhance the relationship between people and the land, while technological advancement ought to support sustainable development and overcome space constraints. Policymakers notice opportunities to use digital technology to improve methods for assessing the environmental impact of urbanisation, local public transportation, digital tourism, water and sewage network planning, energy consumption optimisation, carbon-free investment development, and the prevention of disasters. They recognise the need to incorporate novel infrastructure into spatial planning, such as drone logistics and routes, as well as self-driving car lanes. The authors of the Japanese document emphasise the need to collect geospatial information and data to allow for better organisation of local space, for example for the settlement of abandoned houses and the creation of new attractive and barrier-free public spaces. According to the strategy, it is important to enhance security, assuming that remote monitoring services will cover 150 regions by 2027. Moreover, strategy points to the use of integrated architectural BIM and PLATEAU technologies that allow for the evaluation of the condition and structure of buildings. According to projections, 500 cities will be equipped with 3D models by 2027, supporting city management and space reorganisation. At the same time, policymakers point out that the changes must take into account issues related to education, health care, and social care, and their implementation needs specific mechanisms in the field of cross-sectoral and public–private collaboration.

9.7 Conclusion

Law and politics both connect with and aim to control a changing world. They have the difficulty of modifying their content and impact over time while also reducing hazards that were not considered when decisions were taken. Furthermore, they are tightly linked in any approach that encourages substantial smartness (Hacker and Neyer, 2023). There is no doubt that appropriately shaped public-sector activities may contribute to implementing smart solutions that combine the city's progress with the protection of highly valued assets.

When comparing the systemic solutions adopted in Japan and the European Union, it emerges that while the background for the application of the smart city idea is rooted in the values of responsible, sustainable, and innovative development, there is a considerable difference in the emphasis and depth of the connections between distinct components stating the meaning and potential of this idea within spatial planning. While disparities in the definitions of the smart city visible in government documents directly dedicated to this concept may appear figurative and supplementary to the actual nature of the final projects, the differently shaped legal framework and varying positioning of this idea in development strategies show that both lawmakers and policymakers recognise the scope of its impact on sustainable spatial development in slightly different ways.

The EU directives co-defining the planning process clearly favour innovative solutions in city development, with comprehensive consideration of all the elements determining its sustainable character. There is a visible decoding and strong expression of horizontal and cross-sectoral social, climatic, and environmental connections that constitute the core of the implementation of changes. In contrast, Japanese laws show a noticeable fragmentation of factors related to sustainable development, as well as a vertical restriction of the referring criteria in particular area-oriented laws. This distinction could be attributed to the nature and pragmatics of both types of acts, as well as the dates they were created. Directives are typically broad, have the character of a normative guide, and are implemented secondarily by national legislatures. However, based on this logic, Japanese solutions should be even more precise in the application of all factors. Furthermore, Japanese laws, which are rarely and sparingly amended, were enacted before the concept of sustainable development was developed. Nevertheless, this does not change the fact that they shape the current legal environment and axiological foundation.

Interestingly, this is not the case with strategy. EU policymakers comprehensively and in detail defined the need for both the implementation of innovative technologies and sustainable spatial planning in city development, but they did so vertically, in many, generally interconnected documents, with very few details specifying the connections between individual activities that could integrate both concepts. In contrast, in Japan, a single document thoroughly and holistically describes the function of smart solutions, their role in fulfilling the aims and challenges of sustainable development, and the tools for their application in spatial planning.

It should be noted that the aforementioned method of creating official state papers does not imply that the solutions used in both cases are incorrect or defective. However, they require public authorities to be vigilant about their adequacy and effectiveness. In the legal sphere, particularly in Japan, legislators should monitor their usefulness for technological advances, capacity to achieve sustainable development, and mutual consideration during planning and investment processes. In the sphere of strategy, especially EU policymakers should monitor whether the technological changes implemented in Member States are appropriately correlated with the spatial development of the city and whether the adoption of a coherent EU policy in this area would make them more dynamic and effective. This seems particularly important due to the fact that European-wide collaboration on spatial planning and urban development has provided a forum to achieve a better understanding of the role and principles of spatial planning and important benchmarks against which to compare and stimulate innovation in domestic planning practice (Dühr, Colomb, and Nadin, 2010). Therefore, if member states decide to further Europeanise spatial planning, it seems necessary to integrate and systematise the proposals specified in many strategies.

References

Act on Services Related to Waterways (1950) No. 102. In Japanese: 水路業務法 [suirogyōmuhō].

Act on Special Measures Concerning the Procurement of Electricity from Renewable Energy Sources by Electricity Utilities (2011) No. 108. In Japanese: 電気事業者による

再生可能エネルギー電気の調達に関する特別措置法 [denki jigyō-sha ni yoru saisei kanō enerugī denki no chōtatsu ni kansuru tokubetsu sochi-hō].

Act on Special Measures Concerning the Promotion of New Energy Usage (1997) No. 37. In Japanese: 新エネルギー利用等の促進に関する特別措置法 [shin enerugī riyō-tō no sokushin ni kansuru tokubetsu sochi-hō].

Act on the Rational Use of Energy (1979) No. 49. In Japanese: エネルギーの使用の合理化に関する法律 [enerugī no shiyō no gōri-ka ni kansuru hōritsu].

Ahvenniemi, H. *et al.* (2017) "What Are the Differences Between Sustainable and Smart Cities?", *Cities*, 60, pp. 234–245.

Angelidou, M. (2014) "Smart City Policies: A Spatial Approach", *Cities*, 41, pp. S3–S11.

Anthopoulos, L.G. (2017) *Understanding Smart Cities: A Tool for Smart Government or an Industrial Trick?* Springer: Cham.

Asgarzadeh, M. *et al.* (2010) "Investigating Green Urbanism; Building Oppressiveness", *Journal of Asian Architecture and Building Engineering*, 9(2), pp. 555–562.

Balogun, A.-L. *et al.* (2020) "Assessing the Potentials of Digitalization as a Tool for Climate Change Adaptation and Sustainable Development in Urban Centres", *Sustainable Cities and Society*, 53, p. 101888.

Binbusayyis, A. *et al.* (2024) "A Deep Learning Approach for Prediction of Air Quality Index in Smart City", *Discover Sustainability*, 5(89), pp. 1–23.

Bourdic, L., Salat, S. and Nowacki, C. (2012) "Assessing Cities: A New System of Cross-Scale Spatial Indicators", *Building Research and Information*, 40, pp. 1–14.

Building Standard Act (1950) No. 201. In Japanese: 建築基準法 [kenchiku kijun-hō].

City Planning Act (1968) No. 100. In Japanese: 都市計画法 [toshi keikaku-hō].

Correia, D., Marques, J.L. and Teixeira, L. (2022) "The State-of-the-Art of Smart Cities in the European Union", *Smart Cities*, 5, pp. 1776–1810.

Correia, L.M. and Wünstel, K. (2011) "Smart Cities Applications and Requirements. White Paper of the Experts Working Group", Net!Works European Technology Platform.

Council Directive 1999/31/EC (1999) Council Directive of 26 April 1999 on the landfill of waste, O.J. L 182.

Degórska, B. (2021) "Impact of European Union Law and Policy on Spatial Planning and Territorial Governance of the Member States as Regards Selected Environmental Issues", *Przegląd Geograficzny* [Geographical Review], 93(4), pp. 555–585.

Directive 2000/60/EC (2000) Directive of the European Parliament and of the Council of 23 October 2000 Establishing a Framework for Community Action in the Field of Water Policy, O.J. L 327.

Directive 2001/42/EC (2001) Directive of the European Parliament and of the Council of 27 June 2001 on the Assessment of the Effects of Certain Plans and Programmes on the Environment, O.J. L 197.

Directive 2007/2/EC (2007) of the European Parliament and of the Council of 14 March 2007 establishing an Infrastructure for Spatial Information in the European Community (INSPIRE), O.J. L 108.

Directive 2007/60/EC (2007) Directive of the European Parliament and of the Council of 23 October 2007 on the Assessment and Management of Flood Risks, O.J. L 288.

Directive 2008/98/EC (2008) Directive of the European Parliament and of the Council of 19 November 2008 on Waste and Repealing Certain Directives, O.J. L 312.

Directive 2012/27/EU (2012) Directive of the European Parliament and of the Council of 25 October 2012 on Energy Efficiency, Amending Directives 2009/125/EC and 2010/30/EU and Repealing Directives 2004/8/EC and 2006/32/EC, O.J. L 315.

Directive 2018/2001 (2018) Directive of the European Parliament and of the Council of 11 December 2018 on the Promotion of the Use of Energy from Renewable Sources, O.J. L 328.

Dühr, S., Colomb, C. and Nadin, V. (2010) *European Spatial Planning and Territorial Cooperation*. London: Routledge.

Environmental Impact Assessment Act (1997) No. 81. In Japanese: 環境影響評価法 [kankyō eikyō hyōka-hō].

ESPON (2018) "COMPASS – Comparative Analysis of Territorial Governance and Spatial Planning Systems in Europe: Final Report". Available at: www.espon.eu/sites/default/files/attachments/1.%20COMPASS_Final_Report.pdf (Accessed: 14 May 2024).

EU–Japan Center for Industrial Cooperation (2014) "Smart Cities in Japan: An Assessment on the Potential for EU-Japan Cooperation and Business Development". Available at: www.eu-japan.eu/sites/default/files/publications/docs/smartcityjapan.pdf (Accessed: 14 April 2024).

European Commission (2019) "Main Land-Use Patterns in the EU Within 2015–2030". Available at: https://joint-research-centre.ec.europa.eu/reports-and-technical-documentation/main-land-use-patterns-eu-within-2015-2030_en (Accessed: 5 May 2024).

European Commission (2021) "EU–Japan Joint Study on Demographic Trends and Territorial Policy Responses". Available at: https://ec.europa.eu/regional_policy/en/information/publications/studies/2021/eu-japan-joint-study-on-demographic-trends-and-territorial-policy-responses (Accessed: 1 May 2024).

European Commission (2024) "Smart Cities: Cities Using Technological Solutions to Improve the Management and Efficiency of the Urban Environment". Available at: https://commission.europa.eu/eu-regional-and-urban-development/topics/cities-and-urban-development/city-initiatives/smart-cities_en (Accessed: 14 April 2024).

European Committee of the Regions, Commission for Territorial Cohesion Policy and EU Budget, Dallhammer, E. *et al.* (2018) "Spatial Planning and Governance within EU Policies and Legislation and Their Relevance to the New Urban Agenda". Available at: https://data.europa.eu/doi/10.2863/0251 (Accessed: 14 May 2024).

Foth, M. and Caldwell, G. (2018) "More-than-Human Media Architecture", in *MAB18: Proceedings of the 4th Media Architecture Biennale Conference*, pp. 66–75.

Hacker, P. and Neyer, J. (2023) "Substantively Smart Cities – Participation, Fundamental Rights and Temporality", *Internet Policy Review*, 12(1), pp. 1–30.

IPCC (2022) "Climate Change 2022: Impacts, Adaptation and Vulnerability". Available at: https://report.ipcc.ch/ar6/wg2/IPCC_AR6_WGII_FullReport.pdf (Accessed: 14 May 2024).

Mora, H. *et al.* (2021) "Blockchain Technologies to Address Smart City and Society Challenges", *Computers in Human Behavior*, 122, p. 106854.

National Land Use Planning Act (1974) No. 92. In Japanese: 国土利用計画法 [kokudo riyō keikaku-hō].

National Spatial Planning Act (1950) No. 205. In Japanese: 国土形成計画法 [kokudo keisei keikaku-hō].

National Spatial Strategy (2023). In Japanese: 第三次国土形成計画 [dai sanji kokudo keisei keikaku].

Nilon, C., Berkowitz, A., and Hollweg, K. (1999) "Understanding Urban Ecosystems: A New Frontier for Science and Education", *Urban Ecosystems*, 3, pp. 3–4.

Parmentola, A. *et al.* (2022) "Is Blockchain Able to Enhance Environmental Sustainability? A Systematic Review and Research Agenda from the Perspective of Sustainable Development Goals (SDGs)", *Business Strategy and the Environment*, 31(1), pp. 194–217.

Prime Minister's Office of Japan (2021) "Japan's Smart Cities: Main Report". Available at: www.kantei.go.jp/jp/singi/keikyou/pdf/Japan's_Smart_Cities-1(Main_Report).pdf (Accessed: 14 April 2024).

River Act (1964) No. 167. In Japanese: 河川法 [kasen-hō].

Santos, T. *et al.* (2021) "Assessing Sustainable Urban Development Trends in a Dynamic Tourist Coastal Area Using 3D Spatial Indicators", *Energies*, 14, p. 5044.

Schraven, D., Joss, S. and de Jong, M. (2021) "Past, Present, Future: Engagement with Sustainable Urban Development Through 35 City Labels in the Scientific Literature 1990–2019", *Journal of Cleaner Production*, 292, p. 125924.

Sharifi, A. (2019) "A Critical Review of Selected Smart City Assessment Tools and Indicator Sets", *Journal of Cleaner Production*, 233, pp. 1269–1283.

Sharifi, A. *et al.* (2024) "Smart Cities and Sustainable Development Goals (SDGs): A Systematic Literature Review of Co-Benefits and Trade-Offs", *Cities*, 146, p. 104659.

Soe, R.M. (2020) "Mobility in Smart Cities: Will Automated Vehicles Take It Over?", in N. Lopes (ed.), *Smart Governance for Cities: Perspectives and Experiences. EAI/Springer Innovations in Communication and Computing*. Cham: Springer.

Sokołowski, M.M. (2022) *Energy Transition of the Electricity Sectors in the European Union and Japan: Regulatory Models and Legislative Solutions*. Cham: Palgrave Macmillan.

Sokołowski, M.M. and Kurokawa, S. (2022) "Energy Justice in Japan's Energy Transition: Pillars of Just 2050 Carbon Neutrality", *Journal of World Energy Law & Business*, 15(3), pp. 183–192.

Sokołowski, M.M. and Visvizi, A. (eds) (2023) *Routledge Handbook of Energy Communities and Smart Cities*. London: Routledge.

Tikhaleva, E. (2023). "'Smart Cities': Legal Regulation and Potential of Development", *Journal of Digital Technologies and Law*, 1, pp. 803–824.

Tito, M. *et al.* (2023) "Energy Efficiency as a Tool for Energy Poverty Reduction: A Legal Perspective", in M.M. Sokołowski and A. Visvizi (eds), *Routledge Handbook of Energy Communities and Smart Cities*. London: Routledge, pp. 155–171.

UN-Habitat (2022) "Annual Report 2022". Available at: https://unhabitat.org/annual-report-2022 (Accessed: 14 May 2024).

Urban Renewal Act (1969) No. 38. In Japanese: 都市計画法 [toshi keikaku-hō].

Water Pollution Prevention Act (1970) No. 138. In Japanese: 水質汚濁防止法 [suishitsu odaku bōshi-hō].

Yigitcanlar, T., Foth, M. and Kamruzzaman, M. (2019) "Towards Post-Anthropocentric Cities: Reconceptualizing Smart Cities to Evade Urban Ecocide", *Journal of Urban Technology*, 26(2), pp. 147–152.

Yigitcanlar, T. and Kamruzzaman, M. (2018) "Does Smart City Policy Lead to Sustainability of Cities?", *Land Use Policy*, 73, pp. 49–58.

10 Australia Powering Japan's Energy Transition

Madeline Taylor

10.1 Introduction

Decarbonisation is the essential pillar to creating smart cities and ultimately a sustainable society. Cities represent 70% of global CO_2 emissions which stem from energy consumption (Barrett *et al.*, 2024). The energy transition requires policy and regulatory settings at the city-level to create resilient, sustainable, functional, and decarbonised urban centres (Sokołowski and Visvizi, 2023). There are currently 231 global standards identified as contributing to the United Nations Sustainable Development Goal 11 to create Sustainable Cities and Communities. These standards largely focus on creating energy communities, deploying innovative decarbonisation technologies supported by decentralised systems, and energy efficiency measures. In both energy-importer and energy-exporter countries alike, the design and execution of smart cities policy present the requirement to consider multiple energy technologies to replace carbon-intensive centralised and synchronous fuel and electricity (Willis and Aurigi, 2020). While Australia and Japan present differences in topography, natural resource availability, and population density, both are exploring the use of hydrogen blending in natural gas networks and rooftop solar photovoltaics embedded in prosumer and peer-to-peer (P2P) trading options in cities (Sokołowski, 2022).

Alongside their shared sights on hydrogen and solar energy technology, Australia is also the most important energy trading partner for Japan. Australia's geographic proximity and natural resource wealth have created the ideal conditions for ongoing energy trade underpinned by Japan's overarching energy policy to attain energy security (Kurokawa, 2023; Ohta and Barrett, 2023). Australia currently supplies 43% of Japan's LNG and 66% of Japan's coal imports (Japan Embassy in Australia, 2022). For more than three decades, Australia has played a pivotal role in shaping Japan's domestic energy landscape. From the first LNG export to Japan in 1989 to the inaugural liquid hydrogen shipment aboard, the "Suiso Frontier" from Australia to Japan in 2022 (Hydrogen Council, 2022), energy trading activities between both countries are characterised by continuous innovation and advancements in energy technology.

DOI: 10.4324/9781003471417-10

The pioneering and longstanding trade relationship between Japan and Australia is marked by cutting-edge innovation and technological breakthroughs. The urgent need to reach net-zero emissions by 2050 for both countries has created the conditions for this energy trade relationship to shift and transform once again. This transition entails a departure from traditional carbon-intensive energy source exports from Australia to Japan such as coal and LNG towards embracing a new era that centres around hydrogen as a versatile decarbonised energy carrier and rooftop solar photovoltaics adoption in cities. The adoption of hydrogen and solar photovoltaics aims to empower urban consumers in smart cities and facilitate their adoption of decarbonised energy for their fuel and electrification needs.

Australia's experience in establishing an emerging regulatory framework to blend hydrogen in existing gas networks presents important lessons for Japan in its ambitions to create a hydrogen-based society (Wen and Aziz, 2024). Australia also presents important lessons in creating foundations for prosumer P2P platforms before the introduction of market-based mechanisms and tariffs to support rooftop solar photovoltaics. Australia holds one of the highest penetration rates of 16.3 GW (gigawatts) of rooftop solar photovoltaic energy in the world. The success of this rapid and sustained penetration now represents 26% of total renewable electricity generation in Australia, which is a testament to fiscal incentives and regulatory design. The renewable energy target (RET) created the foundational lessons learned from the Australian energy transition and smart cities context.

This chapter examines the two critical energy policies and regulatory developments in Australia that are likely to shape the realisation of Japan's Smart City goals. First, the development, blending in gas networks, and shipment of hydrogen for adoption in city centres. Second, the importance of rooftop solar photovoltaics and the development of prosumer policies to create future marketplaces to enable P2P trading. These two pivotal case studies examine pathways for fuel and electrical energy decarbonisation and serve as important examples to guide future Japanese initiatives in rooftop solar photovoltaics and hydrogen adoption in smart cities context.

The chapter is structured in four key parts. First, it examines the shift in the Australia–Japan energy export activities against the backdrop of increasing decarbonisation ambitions underpinned by Australia's Climate Change Act of 2022 (Cth) and the shift from coal and gas to creating the conditions for a future energy trade relationship based on hydrogen. Second, it surveys recent policy and regulatory developments in Australia to enable gas blending in cities enabled by recent amendments to its federal natural gas laws. Third, it probes the smart cities Standards Roadmap and introduction of smart cities standards in Australia drawing on several case studies before examining recent developments in enabling P2P and prosumer activities building on the success of Australia's increasing rooftop solar photovoltaic adoption. Finally, the chapter concludes by presenting the potential lessons and tools to enhance Japan's smart cities energy policymaking drawing on the case studies examined in this chapter.

10.2 The New Australia–Japan Energy Export Relationship: From Coal and Gas to Ushering in a New Hydrogen Era

For over 200 years, Australia has maintained its position as a primary energy exporter and critical pillar in Japan's energy security (Government of Japan, 2021) (Taylor, 2021). Australia began exporting coal to India from the port of Newcastle in 1801 and liquefied natural gas exports commencing in 1989 from Karratha in Western Australia bound for Japan under long-term LNG sale agreements. Currently, Australia is the largest global metallurgical coal exporter and second-largest thermal coal exporter with Japan importing 15% of all metallurgical coal and 13% of all thermal coal (Australian Government, Office of the Chief Economist (2023)) produced annually by Australia. Australian energy policy has consequently largely been oriented towards the development and export of coal and gas coupled with the liberalisation of domestic gas and electricity markets from the 1990s (Godden, 2023). The Australian National Electricity Market (NEM) created a cooperative federalism approach under the National Electricity Law 1996 and the East Coast Gas Market under the National Gas Act (2008) between the federal, state, and territory regulation of energy to provide stabilisation and modulation in the energy system.

The liberalisation of energy markets in Australia underpins its energy export policy domestically, while the Australian–Japanese energy trade relationship has to date been dominated by energy security. Energy security regulation and policy attract a myriad of contrasting definitions as a "polysemic" concept (von Hippel *et al.*, 2011). Energy may include both electricity and raw commodities such as oil, gas, and coal. However, in the context of Australian energy exports to Japan, energy security can be defined as balancing both domestic energy self-reliance and procuring energy commodity exports (Soliman Hunter and Taylor, 2021).

Japan has one of the lowest OECD rates of energy self-sufficiency and energy security. Its self-sufficiency rate is currently 12.1% which is a decrease from 20.2% since the Great East Japan Earthquake in 2011. Against the backdrop of increasingly volatile oil supply and prices which saw West Texas Intermediate (WTI) crude reach 139.130 USD per barrel in March 2022 (Zhang *et al.*, 2023), Australia is well-positioned as a geographically proximate and reliable energy trade partner to support Japan's goals of decarbonisation and energy security.

Legislative targets to reach net-zero emissions by 2050 pursuant to the Climate Change Act (2021, Cth) have created the environment for Australia's most fundamental energy policy change since the liberalisation of its electricity and gas markets. Two key energy policy shifts are evident (Sokołowski and Heffron, 2021). First, in stark contrast to the centralisation, unbundling, and synchronisation of electricity by establishing the NEM, Australia is currently creating planning and commercial regulatory frameworks to support the creation of renewable energy zones (REZ). The NEM is currently comprised of 35.9% renewable energy (Clean Energy Council, 2023). To capitalise on Australia's rich endowment of renewable energy resources to produce decarbonised metals, minerals, and hydrogen export potential, REZs will create decentralised renewable energy and storage hubs largely in rural areas to support Australia's federal targets of reaching 82% renewable energy by 2030.

Second, the renewed Australian Capacity Investment Scheme system is based on competitive tender bids and a government revenue underwriting similar to contracts for difference systems to deliver an additional 32 GW of new renewable and dispatchable capacity by 2030 (Australian Government, 2023). The Capacity Investment Scheme will complement and support investment in REZs as key strategic zones for renewable energy and storage development. The regulation of onshore REZs and priorities areas for offshore wind and hydrogen will create the fundamental decarbonisation generation infrastructure for several sectors. Importantly in the context of Japan, these strategic areas may create the potential export of renewable energy via subsea cabling and renewable and blue hydrogen created using natural gas coupled with carbon capture and storage technologies.

Hydrogen as a versatile energy carrier capable of being produced by a variety of energy inputs may be applied in a variety of end-use applications from transportation to energy storage with varying degrees of commercial viability. Consequently, hydrogen is poised to act as an important pillar in Japan's future energy security strategy to complement and eventually replace reliance on gas for city consumers.

10.3 Enabling Hydrogen in Future Smart Cities: Lessons from Australia

Realising goals to create hydrogen trade and city hydrogen-compatible distribution networks requires retrofitted or new infrastructure and regulation to enable hydrogen blending. From bespoke refuelling hydrogen stations to hydrogen blending standards, hydrogen will require a suite of policy and legislative amendments to reach Japan's desired "hydrogen society" (Burandt, 2021) (The Ministerial Council on Renewable Energy, Hydrogen and Related Issues, 2023). Gas is a staple in Japanese city energy mixes. In Japan, most city gas is natural gas in the form of LNG mostly imported from Australia. For example, in 2021, Tokyo Gas imported 12.59 million tonnes of LNG in FY2020 and sold 13.0 Bcm of city gas to 12.08 million customers (JOGMEC, 2021). Consequently, decarbonising gas is a crucial aspect of Japan's hydrogen and overall energy strategy. However, land constraints and risks of resilience for hydrogen infrastructure in natural disaster events in Japan led to the first Basic Hydrogen Strategy to target the importation of 5–10 Mt of hydrogen by 2050, most of which will be imported from Australia (Ministry of Economy, Trade and Industry of Japan, METI, 2017).

Japan has also actively promoted fuel-cell hydrogen vehicles, the replacement of coal- and gas-fired power plants with hydrogen-based power generation, and recently, the production of e-methane to produce carbon-neutral synthetic methane imported from the Cooper Basin in South Australia. E-methane is usually produced by harnessing a two-step process. First, renewable hydrogen is produced via renewable electricity harnessing electrolysis to create renewable hydrogen. Second, recycled CO_2 coupled with renewable hydrogen and anaerobic digestion in the presence of a catalyst produces synthetic gas (Nemmour et al., 2023). This permits e-methane to harness CO_2 from industrial processes as a natural gas substitute.

Overall, Japan's gas industry is aiming to replace 90% of its city gas supply with e-methane by 2050 (Tokyo Gas, 2023).

Japan is actively exploring the use of hydrogen production to supply cities. For example, the Clean Energy Ministerial Hydrogen Initiative, the H2 Twin Cities programme, seeks to disseminate information and awareness to support local governments to adopt hydrogen-related facilities (Clean Energy Ministerial, 2022). The Fukushima government is also advancing the hydrogen-based society model within its Fukushima Plan for a New Energy Society. The Fukushima Hydrogen Energy Research Field (FH2R) commenced operations in 2020 with a 10,000 kW hydrogen production facility to produce renewable hydrogen from solar photovoltaic panels located at the facility (METI, 2021). The Tokyo metropolitan government has harnessed hydrogen produced in Yamanashi prefecture to supply the Tokyo Big Sight governmental facility.

The CC Kawasaki Energy Park Facility is also targeting the production of renewable hydrogen harnessing its existing renewable energy facilities to potentially supply hydrogen to Narita Airport (CC Kawasaki Energy Park, 2024). However, currently there are no specific hydrogen laws in Japan. Rather, existing gas regulation in the High Pressure Gas Safety Act (1951) applies to hydrogen by default. In order to support the large-scale blending of hydrogen in city gas networks and the potential distribution of e-methane and hydrogen requires regulation to support production, safety, and distribution in Japan. Australia's 130 hydrogen- and ammonia-related production projects may provide crucial recommendations for regulatory reform in encouraging hydrogen blending and e-methane in existing city gas networks as examined later.

10.3.1 Hydrogen Blending Laws

Australia's National Hydrogen Strategy (Australian Hydrogen Strategy) was released in 2019 and is currently under review with an expected strategy update in late 2024. The current 2019 Australian Hydrogen Strategy has focused on Australia's goal to produce a clean hydrogen supply chain and be among the top three suppliers of hydrogen to the Asian market by 2030 (COAG Energy Council, 2019). The clear focus of the Australian Hydrogen Strategy is to build domestic hydrogen production capacity for export (Salehi *et al.*, 2022). An initial project under the six works stream endorsed under the Australian Hydrogen Strategy is to blend up to 10% hydrogen in domestic natural gas networks. Domestic hydrogen networks will create the foundational transportation and distribution networks for hydrogen in cities and support future hydrogen export hubs.

Blending hydrogen in up to 10% natural gas networks also directly adopts the IEA recommendation to build on existing gas infrastructure to "springboard" hydrogen uptake (Taylor and Soliman Hunter, 2019) and inject hydrogen in town gas networks as evident in Singapore, Hong Kong, and Hawaii (IEA, 2023). For example, the ATCO Community Blending Project in Cockburn, Western Australia, is one of the largest hydrogen blending projects globally. The project blends 2–5%

renewable hydrogen in gas networks to support 27,000 customers (CSIRO, 2023). The blending project commenced in 2021 and has successfully blended hydrogen to supplement rooftop solar energy in the Cockburn community.

To enable gas blending in cities building off the success of the ATCO Community Blending Project, Australia has recently amended its gas laws to incorporate hydrogen. The transportation of gas in Australian jurisdictions is governed by the National Gas Law (NGL) and The National Gas Rules (NGR). The Natural Gas Objective (NGO) underpins the NGL policy and is applied and mirrored in all participating East Coast Gas Market states and territories (National Electricity (South Australia) Act (1996) SA).[1] Previously, the NGL applied to "natural gas" which was defined as a substance in "a gaseous state at standard temperature and pressure; and consists of naturally occurring hydrocarbons, or a naturally occurring mixture of hydrocarbons and non-hydrocarbons, the principal constituent of which is methane".[2] This limitation to naturally occurring hydrocarbons prevented hydrogen blending in city gas network pipelines, as hydrogen requires processing.

Following the passage of the National Energy Laws (Other Gases) Bill 2023 (SA), the NGL was updated to regulate "covered gases" including natural gas, hydrogen, biomethane, synthetic methane, and blends of these gases. These covered gases are then further regulated into "natural gas equivalents", whereby gas is suitable for use as a natural gas substitute and (a) the gas has been prescribed by a local instrument for use in a jurisdiction or a specified area in the jurisdiction as a natural gas equivalent; or (b) the gas is supplied through an existing distribution system or an extension of an existing distribution system.[3] These natural gas equivalents ensure hydrogen and e-methane can be produced and prescribed by any Australian state and territory, and existing or extensions of existing distribution systems must be made to accommodate this blending. "Prescribed covered gases" are the final category of gases and would cover the distribution of 100% hydrogen or e-methane into cities and other regions. The Australian government is also currently examining the amendments required to enable 100% hydrogen or e-methane distribution by amending the National Energy Retail Regulation 2011 (SA).

The need to rapidly decarbonise gas networks in Japan and Australia has seen regulatory reform focusing initially on enabling hydrogen blending to create foundational definitions and the regulation of hydrogen distribution for use in high energy consumption areas, like cities. Hydrogen is also valuable in its versatility for domestic consumption and to create energy security. However, the electrification of urban centres and households, particularly household appliances and heating and cooling systems, is gaining increasing traction and focus in Australia as a highly efficient and cost-effective mechanism comparable to hydrogen to reach smart city goals. Consequently, the next part of this chapter examines the rise of smart city standards in Australia to reach net-zero emissions in city centres to expose important lessons in smart city policy development for Japan. In particular, it examines the concept of prosumerism sparking a new electricity marketplace to permit the trade and supply of energy between consumers without an active electricity retailer.

10.4 Smart Cities in Australia: Standards Forming the Pillar of Future Decarbonisation

The concept of smart cities elicits a wide array of diverse and multifaceted definitions. In the Australian context, the Australian government defines smart cities as "those which apply innovative technologies to enhance urban services, reduce costs and resource consumption, and to engage more effectively with citizens" (Australian Parliament, 2018, 41). The important intersection between decarbonisation and creating citizen-centric technology is inherently linked with smart cities in Australia. Australia's population is projected to reach 30 million in 2033 from 26.63 million in 2023 (ABS, 2024), and 90% of the Australian population lives in cities. Most of this population growth and associated energy consumption and emissions will be based in Australia's top three cities: Sydney, Melbourne, and Brisbane.

The need to embed smart technologies to reach net-zero emissions goals in cities is paramount in Australia for both electricity system resilience and energy affordability. The NEM has experienced several volatility and reliability events including electricity prices exceeding 5,000 AUD per megawatt-hour (MWh) in the NEM from the July to September 2023 quarter (Austrailan Energy Regulator, 2023) and historically poor levels of reliability for an ageing thermal (coal and gas) generation fleet with generators withdrawing during global gas and coal price spikes and scarcity in 2022 and 2023 (AEMO, 2023). Creating renewable and efficient energy infrastructure is critical to reaching and increasing city-level decarbonisation ambitions. For example, in the city of Sydney, all city buildings and streetlights are powered entirely by renewable energy under three Power Purchase Agreements based in regional NSW from 2020. Finally, Sydney aims to be a net-zero emissions city by 2035 (City of Sydney, 2021).

Australia's first Smart Cities policy was created in 2016 in the Smart Cities Plan underpinned by a 50 million AUD grants subsidisation scheme between 2017 and 2020. The program supported collaborative projects to improve the liveability and sustainability of cities and suburbs, increase public and private data sets to support citizen engagement, increase innovation and capability in local governments, and contribute to the development of Smart City standards and improvement of regulation (Australian Government, 2018). Several technologies incorporated into 52 projects were funded in the 2017 first round of grants including the Switched on Darwin project which received a 5 million AUD grant to upgrade 684 CBD streetlights to LED smart lighting which can be controlled through technology to adjust the lighting level to respond to activity in the city (City of Darwin, 2019).

The regulation of smart cities in Australia represents a patchwork of state-level and industry-led building standards, technology laws, energy efficiency laws, commercial energy supply agreements, and planning and development for net-zero energy technology infrastructure (Hu, 2020). This regulatory complexity led to the development of The Building Up & Moving Out Report produced in 2018 as a recommendation of a result of the federal senate inquiry into the Australian government's role in the development of cities (The Inquiry, Parliament of

Australia, 2018). The inquiry outlined a 50-year roadmap to create sustainability transitions in existing cities, and second, to grow new and transition existing sustainable regional cities and towns. Reduction in energy consumption to enhance liveability and quality of life was one of the key focus areas of The Inquiry which produced 37 recommendations including Recommendation 18 which required the creation of a Smart Cities Standards Roadmap.

Despite the success of the Smart Cities Plan, grants scheme, and overarching Building Up & Moving Out Framework, the federal Australian policy on smart cities is currently in a state of overhaul. The change of federal government to the Labor Party in 2020 has seen the development of the federal cities and suburbs unit to develop a new broader National Urban Policy framework ranging from climate impacts and decarbonisation from equitable access to jobs, homes, and services to develop a sustainable growth plan for cities. The Urban Policy Forum is currently providing advice on policy development and the first updated State of Cities Report is expected in 2024. Whether the application of smart energy technologies, similar to the Smart Cities Plan, remains a focus point of the new National Urban Policy remains to be seen. Inevitably, such reform in urban policy requires considerable coordination of state and city differences across Australia.

Uniformity, consistency, and certainty for communities are crucial prior to the development of a dedicated smart cities regulatory framework. Such a framework should not only support the decarbonisation of city infrastructure but also permit increasingly smart consumer-level technologies to coordinate and manage the distribution of renewable energy and the increasing penetration of rooftop solar photovoltaics. In the absence of a streamlined and specific smart cities regulatory framework, Standards Australia has developed a Smart Cities Standards Roadmap to "identify the standards required in each sector to unlock the benefits of connected Australian cities" (Australian Parliament, 2018, p. 29). Such standards may become binding when ratified and adopted into legislation. However, the majority act as guidance for city policymakers.

10.4.1 The Smart Cities Standards Roadmap

The Smart Cities Standards Roadmap importantly identifies the ISO 37100 international series of standards to guide cities to develop their own goals and objectives to develop smart city strategies (Standards Australia, 2020). Interestingly, Standards Australia largely recommends the adoption of the ISO international standards over the creation of Australian standards. This is because Australia is a relatively late adopter of smart cities policies compared with other Asia-Pacific countries, such as Singapore, which produced its first international smart sustainable cities standards case study and report in 2017 (ITU, 2017). Currently, Standards Australia has released detailed guidance on three international ISO 27100 standards, the most important of which are: ISO 37155-1:2020 "Framework for integration and operation of smart community infrastructures – Part 1: Recommendations for considering opportunities and challenges from interactions in smart community infrastructures from relevant aspects through the life cycle", and ISO 37155-2:2021 "Framework for integration and operation of smart community infrastructures – Part 2: Holistic approach and the strategy for development, operation and maintenance of smart community infrastructures". Both ISO standards serve as guiding methodologies for city authorities to adopt and encourage smart community infrastructure, such as solar photovoltaic microgrids, to contribute to the "sustainable development and resilience of the community" (Standards Australia, 2020).

Transport for NSW (TfNSW) is the leading state-based transportation provider in New South Wales and has provided one of the first cases of clear methodological adoption of the recommended smart cities standards. In applying the baselines and recommended standards in the Smart Cities Standards Roadmaps, for smart infrastructure development Transport for NSW suggests a four-step process (NSW Transport, Smart Places Playbook Standards, 2022). First, a baseline and maturity assessment for the relevant community is needed (e.g. maturity assessment, ISO 37153); second, guidelines and measures must be adopted (e.g. smart infrastructure, ISO 37155-1); consideration of relevant processes (e.g. security and resilience standards, ISO 22396), and finally relevant technical requirements including whether asset undergrounding is needed (AS 5488). Such a process will be critically important to support the growth of smart grids and smart homes and communities to measure energy consumption and distributed community energy projects.

Building upon the Smart Cities Standards Roadmap, in 2023 Standards Australia's Smart Cities Systems Committee developed the Australian-specific Smart City Standards Inventory and Mapping Methodology within AS IEC SRD 63233.1:2023 (2023, the methodology). The methodology seeks to provide a methodological mapping and smart city reference standards and templates to implement best practice governance, design, development, operation, management, and maintenance. The methodology represents a pivotal first step to developing streamlined standards and best practice smart city ecosystems in Australia. By providing a single source of collated relevant ISO and IEC standards for cities to consider in their smart designs, particularly in relation to renewable energy, the methodology will likely become a critical feature in sustainable and decarbonised cities and communities in the future. To ensure consistent implementation of standards and methodologies ,while accommodating the diverse needs of consumers and communities, a smart cities regulatory framework is needed at the state level in Australia. A state-level regulatory framework may guide city policymakers and enable monitoring of smart city initiatives without mandating specific energy technologies.

10.4.2 Rooftop Solar Photovoltaics and Prosumerism in Australia: The Rise of Smart City Consumers

Since 2019, Australia has held the title as the country with the highest solar photovoltaic installation rate per capita in the world. Australia holds over 1.1 kW of solar power per person and contributes about 25.8% of total Australian renewable energy generation (Clean Energy Council, 2023). This high rooftop solar photovoltaic penetration is a result of the previous RET scheme which commenced in 2001 and will be phased out in 2030. The RET scheme was split into large- and small-scale schemes to incentivise the uptake of renewable energy pursuant to the Renewable Energy (Electricity) Act 2000 (Cth). The small-scale renewable energy scheme incentivises the installation of rooftop solar photovoltaics by the award of small-scale technology certificates (STCs) for each megawatt hour of renewable energy generated or displaced by the solar system. Liable entities defined under Section 35 of the RET scheme of Renewable Energy (Electricity) Act 2000 (Cth) hold a legal obligation to buy and surrender such STCs to the Clean Energy Regulator and the REC Registry each quarter establishing a tradable marketplace to stimulate rooftop solar photovoltaic uptake (Soliman Hunter and Taylor, 2023).

The RET system created the conditions for consumers and entities to invest in renewable energy and Power Purchase Agreements at scale. For example, since 2020, the Australian Capital Territory and the city of Canberra have been powered by 100% renewable energy largely underpinned by rooftop solar and Power Purchase Agreements with the Mugga Lane Solar Park, Williamsdale Solar Farm, and Royalla Solar Farm (ACT Government, 2023). As states and cities continue to switch to increased penetration of renewable energy, the challenges of a rapidly changing electricity generation mix and demand profile have caused challenges for the Australian Energy Market Operator (AEMO) to ensure power system flexibility and power system reliability standards enshrined in law. The current reliability standard for the NEM requires 99.998% of forecast customer demand to be met each year (AER, 2019). The shift to rooftop solar photovoltaic energy has caused some concern to ensure reliability in the NEM system where rooftop systems are adopted without household battery storage. As a result, the shift to a regulated prosumerism system will require a pivot from traditional incentivisation and innovative approaches to network capacity building (Kallies, 2023).

Smart technologies alongside the rapid uptake of solar photovoltaics have been comparatively minimal in Australia. The lack of storage, smart meters, and other smart technology has caused significant disruptions to the NEM in the maintenance of grid stability due to the influx of solar photovoltaic causing wholesale electricity prices to drop below zero in the September 2023 quarter by 19% of the time (AER, 2023). Negative pricing events cause coal-fired power plants to ramp down capacity causing significant costs and potential volatility when demand peaks and firmed renewable energy with storage is not available. Consequently, the need to rapidly produce legal frameworks to encourage smart technology uptake and create a prosumer market is urgently needed in Australia.

Prosumers are defined by Wilkinson *et al.* as "actors who both produce and consume electricity" (Wilkinson *et al.*, 2021, p. 1). By 2027, 40% of Australian customers are forecast to become prosumers by harnessing distributed energy resources consisting of rooftop solar photovoltaics, behind-the-metre battery storage, and electric vehicles to generate and consume at their premises (ARENA, 2023). Prosumers are the critical nexus to bind multidirectional electricity flows as a result of increasing rooftop solar photovoltaic uptake, particularly in urban areas, in the shift away from traditional unidirectional energy supply modes. Smart metres and tariff redesign are critical elements to allowing prosumer P2P to trade and negotiate electricity generation and export while acting within network constraints including acceptable voltage ranges defined in the National Electricity Rules (Best, Chareusny and Taylor, 2023).

One of the first P2P systems enabled by blockchain-backed technologies was deployed in the "Renewable Energy and Water Nexus" (RENeW Nexus) project in Fremantle, Western Australia. The RENeW Nexus project produced a prosumer network comprising 50,000 transactions per 1 MWh of P2P energy transacted each month (Powerledger, 2023). Blockchain-enabled smart contracts may provide the backdrop of contractual transactions with consumers removing the need for electricity retailers. P2P models have also been cited as promoting energy justice (Sokołowski and Taylor, 2023) to improve access to affordable renewable energy sources for apartments and buildings providing decarbonisation options for consumers who do not live in single dwelling homes (Wilkinson *et al.*, 2021, p. 2).

Following the success of the RENeW project, questions remain as to the compatibility of current regulatory frameworks to enable a large-scale prosumer marketplace in Australia. While smart contracts are permitted and valid where essential elements are met pursuant to the Electronic Transactions Act (1999, Cth) and mirrored legislation at the state and territory to create legally binding contractual obligations, the P2P model is currently largely unsupported in the Australian National Electricity Rules (Wilkinson *et al.*, 2021, p. 2). Consequently, the Australian Energy Market Commission is currently consulting on a rule change to enable consumer energy resources to be separately identified enabling consumers to "create new secondary settlement points for consumer energy resources behind consumers' current meters" (AEMC, 2022). Such a rule change would be the fundamental first stage in creating P2P-enabled rules to permit consumers to trade and sell directly and provide the enabling regulatory framework for large-scale prosumerism in Australia critical to achieving smart city goals.

This section has provided an overview of key developments in smart cities standards and prosumers in Australia producing new requirements in fiscal incentive redesign, system governance and regulation, and reliability measures to create a new consumer-centric marketplace. It charts the important regulatory developments underpinning Australia's ongoing success in rooftop solar photovoltaic energy in urban centres bringing new challenges in system reliability and the need for increasingly smart technologies to remove energy market negative pricing events.

While the Australian RET regimen represents a highly effective market-based system to generate rooftop solar photovoltaic uptake at scale, future policy focus will firmly remain on incentivising demand-side consumer solar photovoltaic generation and creating contractual marker-based structures. Such structures represent important insights for the Japanese smart city and specifically rooftop solar policies to empower consumers with market-based regulatory protections and incentives to balance and firm electricity supply during periods of high solar energy generation (Iria and Soares, 2023). Prosumerism holds the potential to empower consumers and potentially deliver energy justice for urban consumers in apartment complexes without sole ownership of rooftop space (Kubli, 2018). The significance of regulation integrating smart city energy technologies, including prosumer mechanisms and incentives, is set to increase significantly. This trend is driven by the growing imperative to achieve increasingly ambitious efficiency and sustainability targets for urban areas in both Australia and Japan.

10.5 Conclusion

The energy transition requires distinct policy and regulatory change while rapidly innovating to empower consumers and decarbonising existing fuel networks tailored for urban environments. This chapter has examined two crucial features of creating net-zero-emission smart cities in Australia – hydrogen and rooftop solar photovoltaics. Clear, consistent, and compatible regulation and policies to govern the production and blending of renewable and low-carbon hydrogen are critical to paving the necessary pathways for future hydrogen export from Australia to Japan and domestic distribution of hydrogen in cities. Decentralised and technology-enabled rooftop solar photovoltaic generation in urban areas to support prosumerism trade will play an increasingly important role in encouraging a sustainable and rapid energy transition while mitigating market-based impacts and pricing events. Additionally, international standards to create toolkits for local city councils to adopt Smart Cities Communities policies have proven to be important in Australia to create city-level action supporting renewable energy and potentially hydrogen.

Achieving smart city outcomes starts with energy sources. Australia's export of decarbonised hydrogen and policy learnings on rooftop solar systems is likely to be the hallmark of the evolving and ever-innovative energy trade relationship with Japan. Yet, there is no single silver bullet for success in creating smart cities. Similarly, there is no single energy source and regulatory formula to support the roll-out of hydrogen and solar technologies in cities. For example, as surveyed within this chapter, the RET system, while very successful in generating rooftop solar in Australia, has exposed reliability constraints and pricing spikes creating market uncertainty for regulators and electricity market participants alike. The combination of supportive fiscal measures coupled with clear regulatory guidelines is critical but not a panacea for decarbonised cities in Australia or Japan.

As Australia and Japan work towards ushering in a variety of decarbonised fuel and electricity sources at a rapid scale and pace, energy policy and regulation will continue to adapt to the new market and consumer needs. However, the vital energy trade relationship between Australia and Japan built on decades of exchange and policy learnings will remain. Fostering mutual decarbonisation, Australia will continue to produce and export energy, while Japan seeks to distribute and import decarbonized energy. This shift towards low-carbon imports and technology adoption underscores the importance of integrating and future-proofing energy policy and regulatory frameworks for cities.

Acknowledgements

The author would like to thank and acknowledge the deepest gratitude to the lead editor of this book, Associate Professor Maciej M. Sokołowski. The author is grateful for the invaluable conversations and research undertaken during a research exchange between the author and Professor Sokołowski to Keio University, Meiji University, and Waseda University.

Notes

1 Div 1 s 23 of the National Gas (South Australia) Act 2008 (SA).
2 National Gas (South Australia) Act 2008 (SA) Ch 1.
3 National Energy Laws (Other Gases) Bill 2023 (SA) s 72.

References

ABS (2024) "Population". Available at: www.abs.gov.au/statistics/people/population#:~:text=Statistics%20in%20this%20release%20are,people%20at%2030%20June%202023 (Accessed: 17 January 2024).

ACT Government (2023) "What the ACT Government Is Doing". Available at: www.climatechoices.act.gov.au/energy/what-the-act-government-is-doing (Accessed: 20 January 2024).

AEMC (2022) "Unlocking CER Benefits Through Flexible Trading". Available at: www.aemc.gov.au/sites/default/files/2022-12/Consultation%20paper%20-%20Unlocking%20CER%20benefits.pdf (Accessed: 10 January 2024).

AEMO (2023) "Update to 2022 Electricity Statement of Opportunities". Available at: https://aemo.com.au/-/media/files/electricity/nem/planning_and_forecasting/nem_esoo/2023/february-2023-update-to-the-2022-esoo.pdf?la=en&hash=1AED91846C35DE3DE0BFC071A2228EAD (Accessed: 21 January 2024).

AER (2019) "Values of Customer Reliability". Available at: www.aer.gov.au/industry/registers/resources/reviews/values-customer-reliability-2019 (Accessed: 20 January 2024).

AER (2023) "Electricity Prices Above $5,000 Per MWh July to September". Available at: www.aer.gov.au/system/files/2023-12/AER%20-%20Electricity%20prices%20above%20%245%2C000MWh%20-%20July%20to%20September%202023_0.pdf (Accessed: 5 January 2024).

ARENA (2023) "Future Grid for Distributed Energy ". Available at: https://arena.gov.au/knowledge-bank/future-grid-for-distributed-energy/ (Accessed: 30 March 2024).

AS IEC SRD 63233.1:2023 (2023) "Smart City Standards Inventory and Mapping Part 1: Methodology".

Australian Energy Regulator (2023) "Wholesale Energy Prices and Demand Fall in July to September Period". Available at: www.aer.gov.au/news/articles/news-releases/wholesale-energy-prices-and-demand-fall-july-september-period (Accessed: 18 January 2024).

Australian Government, Department of Climate Change, Energy, the Environment and Water (2023) "About the Capacity Investment Scheme". Available at: www.dcceew.gov.au/energy/renewable/capacity-investment-scheme#:~:text=It%20aims%20to%20help%20build,GW%20of%20capacity%20by%202030 (Accessed: 15 January 2024).

Australian Government, Office of the Chief Economist (2023) "Resources and Energy Quarterly". Available at: www.industry.gov.au/sites/default/files/2023-12/resources-and-energy-quarterly-december-2023.pdf (Accessed: 15 January 2024).

Australian Parliament (2018) "Building Up & Moving Out Inquiry into the Australian Government's Role in the Development of Cities". Available at: https://parlinfo.aph.gov.au/parlInfo/download/committees/reportrep/024151/toc_pdf/BuildingUp&MovingOut.pdf;fileType=application%2Fpdf (Accessed: 17 January 2024).

Barrett, B. *et al.* (2024) "How Can Cities Achieve Accelerated Systemic Decarbonization? Analysis of Six Frontrunner Cities", *Sustainable Cities and Society*, 100, p. 105000.

Best, R., Chareusny, A. and Taylor, M. (2023) "Changes in Inequality for Solar Panel Uptake by Australian Homeowners", *Ecological Economics*, 19, pp. 1–8.

Burandt, T. (2021) "Analyzing the Necessity of Hydrogen Imports for Net-Zero Emission Scenarios in Japan", *Applied Energy*, 298, p. 117265.

CC Kawasaki Energy Park (2024) "Home". Available at: https://eco-miraikan.jp/en/energypark.html (Accessed: 20 January 2024).

City of Darwin (2019) "Switching on Darwin". Available at: www.darwin.nt.gov.au/transforming-darwin/innovation/switching-on-darwin (Accessed: 12 January 2024).

City of Sydney (2021) "Environmental Strategy 2021–2025". Available at: www.cityofsydney.nsw.gov.au/strategies-action-plans/environmental-strategy (Accessed: 10 January 2024).

Clean Energy Council (2023 "Clean Energy Australia Report". Available at: www.cleanenergycouncil.org.au/resources/resources-hub/clean-energy-australia-report (Accessed: 10 December 2023).

Clean Energy Ministerial (2022) "H2 Initiative 2022". Available at: www.cleanenergyministerial.org/h2-initiative-launches-h2-twin-cities-program/ (Accessed: 20 January 2024).

Climate Change Act (2021) Cth.

COAG Energy Council (2019) "Australia's National Hydrogen Strategy". Available at: www.dcceew.gov.au/sites/default/files/documents/australias-national-hydrogen-strategy.pdf (Accessed: 4 January 2024).

CSIRO (2023) "ATCO Hydrogen Blending Project". Available at: https://research.csiro.au/hyresource/atco-hydrogen-blending-project/ (Accessed: 15 January 2024).

Electronic Transactions Act (1999) Cth.

Godden, L. (2023) "Energy Law and Regulation in Australia", in G. Bellantuono *et al.* (eds) *Handbook of Energy Law in the Low-Carbon Transition, Energy Law and Regulation*. Berlin – Boston, MA: De Gruyter, pp. 369–386.

Government of Japan (2021) "Strategic Energy Plan". Available at: www.enecho.meti.go.jp/category/others/basic_plan/pdf/strategic_energy_plan.pdf (Accessed: 12 December 2023).

High Pressure Gas Safety Act (1951).

Hu, R. (2020) "Australia's National Urban Policy: The Smart Cities Agenda in Perspective", *Australian Journal of Social Issues*, 55, pp. 201–217.

Hydrogen Council (2022) "Toward a New Era of Hydrogen Energy: Suiso Frontier Built by Japan's Kawasaki Heavy Industries". Available at: https://hydrogencouncil.com/en/toward-a-new-era-of-hydrogen-energy-suiso-frontier-built-by-japans-kawasaki-heavy-industries/ (Accessed: 20 January 2024).

IEA (2023) "Global Hydrogen Review". Available at: https://iea.blob.core.windows.net/assets/ecdfc3bb-d212-4a4c-9ff7-6ce5b1e19cef/GlobalHydrogenReview2023.pdf (Accessed: 15 January 2024).

Iria, J. and Soares, F. (2023) "An Energy-as-a-Service Business Model for Aggregators of Prosumers", *Applied Energy*, 347, p. 121487.

ISO 37155-1:2020 (2020) "Framework for Integration and Operation of Smart Community Infrastructures – Part 1: Recommendations for Considering Opportunities and Challenges from Interactions in Smart Community Infrastructures from Relevant Aspects Through the Life Cycle".

ISO 37155-2:2021 (2021) "Framework for Integration and Operation of Smart Community Infrastructures – Part 2: Holistic Approach and the Strategy for Development, Operation, and Maintenance of Smart Community Infrastructures".

ITU (2017) "Implementing ITU-T International Standards to Shape Smart Sustainable Cities: The Case of Singapore". Available at: www.itu.int/dms_pub/itu-t/opb/tut/T-TUT-SSCIOT-2017-7-PDF-E.pdf (Accessed: 18 January 2024).

Japan Embassy in Australia (2022) "Australia–Japan Resources and Energy Relationship". Available at: https://japan.embassy.gov.au/tkyo/resources.html#:~:text=On%20the%20flip%20side%2C%20Australia,per%20cent%20thermal)%20in%20CY2022 (Accessed: 19 January 2024).

JOGMEC (2021) "Tokyo Gas". Available at: https://oilgas-info.jogmec.go.jp/_res/projects/default_project/_page_/001/009/258/4_33_Tokyo_gas_2022_en.pdf (Accessed: 10 December 2023).

Kallies, A. (2023) "The Changing Role of Energy Networks in the Energy Transition", in G. Bellantuono *et al.* (eds) *Handbook of Energy Law in the Low-Carbon Transition, Energy Law and Regulation in Australia.* Berlin – Boston. MA: De Gruyter, pp. 203–216.

Kubli, M. (2018) "Squaring the Sunny Circle? On Balancing Distributive Justice of Power Grid Costs and Incentives for Solar Prosumers", *Energy Policy*, 114, pp. 173–188.

Kurokawa, S. (2023) "Energy Law in the Low-Carbon Transition in Japan: The Tough Road to a Low-Carbon Society after the Fukushima Nuclear Crash", in G. Bellantuono *et al.* (eds) *Handbook of Energy Law in the Low-Carbon Transition, Energy Law and Regulation*. Berlin – Boston, MA: De Gruyter, pp. 435–450.

METI (2017) "Basic Hydrogen Strategy Ley Points. Technical Report". Available at: www.meti.go.jp/english/press/2017/pdf/1226_003a.pdf (Accessed: 5 January 2024).

National Electricity (South Australia) Act (1996) SA.

National Energy Laws (Other Gases) Bill (2023) SA.

National Energy Retail Regulation (2011) SA.

National Gas (South Australia) Act (2008) SA.

Nemmour, A., Inayat, A., Janajreh, I. and Ghenai, C. (2023) "Green Hydrogen-Based E-Fuels (E-Methane, E-Methanol, E-Ammonia) to Support Clean Energy Transition: A Literature Review", *International Journal of Hydrogen Energy*, 48, pp. 29011–29033.

NSW Transport (2022) "Smart Places Playbook Standards June 2021". Available at: www.transport.nsw.gov.au/system/files/media/documents/2022/Smart-Places-Playbook-Standards.pdf (Accessed: 18 January 2024).

Ohta, H. and Barrett, B. (2023) "Politics of Climate Change and Energy Policy in Japan: Is Green Transformation Likely?", *Earth System Governance*, 17, p. 100187.

Parliament of Australia (2018) "Building Up & Moving Out". Available at: www.aph.gov.au/Parliamentary_Business/Committees/House/Former_Committees/ITC/DevelopmentofCities/Report (Accessed: 10 January 2024).

Powerledger (2023) "RENeW Nexus, Australian Government, Australia". Available at: www.powerledger.io/clients/renew-nexus-australian-government-australia (Accessed: 20 January 2024).
Renewable Energy (Electricity) Act 2000 (Cth).
Salehi, F. *et al.* (2022) "Overview of Safety Practices in Sustainable Hydrogen Economy – An Australian Perspective", *International Journal of Hydrogen Energy*, 47, p. 34689.
Sokołowski, M.M. (2022) *Energy Transition of the Electricity Sectors in the European Union and Japan Regulatory Models and Legislative Solutions*. Cham: Palgrave Macmillan.
Sokołowski, M.M. and Heffron, R.J. (2021) "Defining and Conceptualising Energy Policy Failure: The When, Where, Why, and How", *Energy Policy*, 161, p. 112745.
Sokołowski, M.M. and Taylor, M. (2023) "Just Energy Business Needed! How to Achieve a Just Energy Transition by Engaging Energy Companies in Reaching Climate Neutrality: (Re)Conceptualising Energy Law for Energy Corporations", *Journal of Energy and Natural Resources Law*, 41, pp. 157–174.
Sokołowski, M.M. and Visvizi, A. (eds) (2023) *Routledge Handbook of Energy Communities and Smart Cities*. London: Routledge.
Soliman Hunter, T. and Taylor, M. (2021) "Long-Term and Short-Term Liquid Fuel Security in Australia – What Role for the Great Australian Bight?", *Energy Policy*, 157, p. 112472.
Soliman Hunter, T. and Taylor, M. (2023) "From Coal to Climate Change: An Australian Perspective on the Energy Transition", in K. Gromek-Broc (ed.), *Regional Approaches to the Energy Transition*. New York, NY: Springer, pp. 245–269.
Standards Australia (2020) "Smart Cities Standards Roadmap". Available at: www.standards.org.au/documents/sa-smart-cities-roadmap (Accessed: 5 January 2024).
Taylor, A. (2021) "Japan–Australia Partnership on Decarbonisation Through Technology". Available at: www.minister.industry.gov.au/ministers/taylor/media-releases/japan-australia-partnership-decarbonisation-through-technology (Accessed: 10 November 2023).
Taylor, M. and Soliman Hunter, T. (2019) "The Hydrogen Hope? Challenges and Opportunities for an Australian Hydrogen Industry", *OGEL*, 19, pp. 1–17.
The Ministerial Council on Renewable Energy, Hydrogen and Related Issues (2023) "Basic Hydrogen Strategy". Available at: www.meti.go.jp/shingikai/enecho/shoene_shinene/suiso_seisaku/pdf/20230606_5.pdf (Accessed: 5 January 2024).
Tokyo Gas (2023) "Commencement of the Joint Feasibility Study with Santos for Production and Export of E-Methane in Australia". Available at: www.tokyo-gas.co.jp/en/IR/support/pdf/20231121-01e.pdf (Accessed: 10 January 2024).
von Hippel, D.F. *et al* (2011) "Evaluating the Energy Security Impacts of Energy Policies", in B. Sovacool (ed.), *The Routledge Handbook of Energy Security*. London: Routledge, pp. 74–95.
Wen, D. and Aziz, M. (2024) "Perspective of Staged Hydrogen Economy in Japan: A Case Study Based on the Data-Driven Method", *Renewable and Sustainable Energy Reviews*, 189, p. 113907.
Wilkinson, S., John, M. and Morrison, G.M. (2021) "Rooftop PV and the Renewable Energy Transition: A Review of Driving Forces and Analytical Frameworks", *Sustainability*, 13(10), p. 5613.
Willis, K. and Aurigi, A. (2020) *The Routledge Companion to Smart Cities*. London: Routledge.
Zhang, Q. *et al.* (2023) "Unveiling the Impact of Geopolitical Conflict on Oil Prices: A Case Study of the Russia–Ukraine War and Its Channels", *Energy Economics*, 126, p. 106956.

11 Low-Carbon Transition and Energy Poverty in a Smart City

Satoshi Kurokawa

11.1 Introduction

An energy transition towards carbon neutrality by 2050 has a potential to mitigate energy poverty, and there are different ways to do it. In Japan, the government has prompted residents to install photovoltaics (PVs) on their rooftops and use the generated electricity. This policy supports renewable energy growth and reduces household electricity expenditures. Households living in zero-energy houses (ZEHs), which have PVs and storage batteries, pay low electricity fees. Battery electric vehicles (BEVs) are charged using electricity from PVs. However, households in poverty, which cannot afford PVs and BEVs, could be left behind in this trend and continue to pay rising electricity bills. This implies that an energy transition in a smart city has the potential to prompt energy poverty. In this context, this chapter elaborates on the energy transition towards carbon neutrality in the urban residential sector of Japan in relation to the issue of energy poverty.

Japan needs to shift energy sources from fossil fuels to zero-emission energy sources, such as renewable and nuclear energy sources, towards carbon neutrality by 2050 (Sokołowski, 2022, p. 103). Because of the 2011 Fukushima nuclear power station accident, expanding the generation capacity of nuclear power stations is challenging. Therefore, the expansion of renewable energy sources has become inevitable in Japan. The output fluctuations of electricity from renewable energy sources must be addressed to obtain more electricity from these sources. Smart houses with PVs and storage batteries can become stabilisers that absorb the output fluctuations of renewable energy electricity in the future.

The government strongly promotes ZEHs as a method of reducing CO_2 emissions from the residential sector. A ZEH is well insulated from the outside temperature and is equipped with a PV and storage battery. It is managed by an intelligent home energy management system (HEMS) and becomes part of the smart grid through a smart metre. The smart grid collects information on electricity demand and controls the electricity supply based on this information. However, in the future, excess electricity is expected to be stored in home storage batteries or BEVs when the electricity supply exceeds the demand. It is also expected that HEMSs will discharge stored electricity to the grid when the electricity supply is short.

DOI: 10.4324/9781003471417-11

BEVs, being no-emission vehicles, are important components of smart cities. They function as storage batteries when connected to a HEMS for charging. BEVs, PVs, and home storage batteries can supply electricity to households when electricity supply from the grid is cut off owing to a disaster. They function as distributed electricity generators in networks and improve the resilience of networks, which is an important consideration in Japan, where natural disasters such as earthquakes often occur.

Electricity prices are increasing during the energy transition towards carbon neutrality, and these rising electricity bills will directly affect the budget of households unless households generate electricity on their rooftops. Low-income households usually live in inexpensive rental housing that is not well insulated or equipped with PVs and storage batteries. Therefore, high electricity bills negatively affect their budgets and can easily cause energy poverty. Owing to economic factors, poor households find it difficult to access the energy necessary to support their lives and, sometimes, to keep their rooms warm. Increasing energy poverty can hinder the energy transition towards carbon neutrality unless the government provides appropriate assistance.

11.2 Green Transformation and Its Impact on Households

The Act on Promotion of Smooth Transition to a Decarbonized Growth-Oriented Economic Structure (GX Promotion Act, 2023) established a programme to prompt an energy transition towards a low-carbon society. In 2021, residences, private vehicles, other transport sectors, and other service sectors emitted 14.7%, 5.1%, 12.3%, and 17.9% of all CO_2 emissions in Japan, respectively (Ministry of the Environment, MoE and National Institute for Environmental Studies, NIES, 2023, p. 17). Therefore, approximately half of Japan's CO_2 emissions are attributed to socio-economic activities in city areas across the residential, commercial (stores and offices), and transport sectors. The City Planning Act for Promotion of Low Carbon Cities (Eco-town Act, 2012) endorsed the creation of energy-efficient compact cities through urban planning. Meanwhile, the GX Promotion Act (2023) has rigorously accelerated the energy transition in city areas towards a low-carbon society.

11.2.1 GX in Residential Areas

In Japan, 92% of the population live in urban areas (United Nations Department of Economic and Social Affairs, 2018, p. 37). The residential sector accounts for approximately 15.1% of the final energy consumption (ANRE, 2023c, p. 3). Therefore, realising net-zero emissions from the residential sector is necessary for carbon neutrality. Electrifying energy and generating electricity from renewable energy sources are major strategies for energy transition in the residential sector. Electricity generation by PVs alone contributes to renewable energy penetration in Japan. Households with PVs will be recognised as distributed electricity generators in smart cities. Electricity from rooftop PVs is consumed directly in a house, charged in the storage battery, including that of a BEV, or sold to electricity companies. If consumed or stored inside a house, the fluctuation in renewable energy electricity does not influence the grid. During the process of energy transition, a HEMS will manage PVs, storage batteries, and IoT appliances, including a hot water tank, and will create an electricity demand or supply, thereby contributing to stabilising the grid electricity quality.

In Japan, the GX policy promotes ZEHs whose energy consumption is net zero. Almost all the energy consumed in a ZEH is generated by a rooftop PV. The electricity charged into the storage battery during the day is used at night. The ZEH is well insulated from the outside cold temperature for efficient heating. Appliances are energy efficient and meet top-runner energy-efficiency standards by a sufficient margin. Therefore, households living in ZEHs pay little to electricity companies. A ZEH resident is paid for excess electricity sold to electricity companies. Thus, a household that can afford to live in a ZEH enjoys the benefits of the energy transition while simultaneously contributing to it. However, a household that cannot afford to live in a house with PV systems must pay the rising electricity bill. An energy transition in a smart city widens the inequality between the rich and poor. Therefore, the government must create a ZEH promotion scheme that helps poor households and avoids this injustice.

Nevertheless, the PV installation level in energy-poor and low-income households is lower than that of households with average and above-average incomes (Chapman and Okushima, 2019, p. 6). Electricity fees for domestic use in Japan have risen partly owing to carbon pricing policies such as feed-in tariffs (FITs) and partly because of electricity supply shortages caused by reduced investment in thermal power plants, in addition to the shutdown of nuclear power plants. Indeed, soaring fossil fuel prices in international markets have influenced price increases in electricity. For example, the monthly electricity bill of households that consumed 500 kWh (50A contract with the Tokyo Electric Power Company – TEPCO) increased from 13,150 JPY in August 2012 to 16,175 JPY in June 2022. The price of renewables surcharge under the FIT scheme was 0.28 JPY in 2012 and 3.45 JPY in 2022. The higher the electricity price, the larger the gap between households with and without PVs. Therefore, low-income households without PV systems suffer from increasing electricity bills more severely than those with PV systems.

The cost of PVs can be a barrier to the energy transition. The average cost of PV facility installation in a new house in 2022 was 0.261 million JPY/kW (METI, 2023, p. 22). If a house with four members installed a 4 kW capacity solar panel, it would cost over 1 million JPY. The energy justice perspective requires governments to prevent this injustice from happening (Sokołowski and Kurokawa, 2022), as energy poverty might become an obstacle to achieving carbon neutrality (Chapman and Okushima, 2019, p. 1).

11.2.2 Electrification with Electricity Generated by Non-Fossil Fuels

To achieve carbon neutrality, electricity, rather than fossil fuels, should be used as an energy source. In residences, gas stoves should be replaced with electric stoves, and the gas hot water systems should be replaced with electric hot water systems. Vehicles should be powered by electricity rather than gasoline. All-electric houses and BEVs are plausible alternatives for smart cities. Importantly, electricity should be generated from zero-emission energy sources such as renewable energy sources. If electricity is generated by fossil fuels combustion, all-electric houses and BEVs will not reduce CO_2 emissions. Currently, in Japan, more than 60% of the final energy consumption is in the form of fossil fuel combustion. The development of electrification will increase the electricity demand, and additional electricity will be generated at thermal power plants. This will decrease the share of renewable energy electricity in all electricity supplies. To avoid this problem, further renewable energy penetration is necessary.

When focusing on residential areas, on average, a household consumed 15 GJ of electricity, 10.7 GJ of gases, and 5.1 GJ of kerosene in FY2021 (MoE, 2023) in Japan.[1] Shifting from gas and kerosene to electricity reduces CO_2 emissions from houses. Generating electricity on rooftops and using it are favourable scenarios. BEVs should be charged with electricity from rooftop PV systems. Surplus electricity that is not used in a house is fed into a grid automatically under the FIT scheme or a contract with an electricity company.

11.2.3 Rise in Electricity Price and Protection of Household Budgets

During renewable energy penetration, electricity bills increase owing to the renewable energy encouragement surcharge added to the electricity bills based on the FIT scheme. In addition, the supply–demand gap during the peak season increases the price of electricity. This gap is caused by the shrinking generation capacity of thermal power plants. Electricity companies are reluctant to invest in thermal power plants because they have become stranded assets. The levelized cost of electricity from fossil fuels is increasing compared with that from renewable energy sources. Therefore, in addition to high fossil fuel prices in international markets, electricity prices for households have increased substantially. Under this circumstance, generating electricity on rooftops is an economically rational behaviour that protects households from expensive electricity prices. Additionally, reducing energy consumption using IoT appliances is effective for reducing electricity bills.

11.2.4 ITMSs and BEVs

The government intends to replace fuel-powered vehicles with EVs. EVs are software-defined vehicles that are expected to be connected to an intelligent traffic management system (ITMS) in a smart city. An ITMS effectively manages traffic and reduces traffic jams. Smooth traffic without congestion reduces the energy consumption of vehicles, resulting in reduced CO_2 emissions. In addition, BEVs can be a type of storage battery for residences and have become a part of HEMSs. In such a case, a BEV is charged inexpensively using the electricity generated by the PV system on a rooftop. BEVs can transfer electricity from residences to other places where electricity is demanded. In this sense, BEVs will become part of community energy management systems. BEV owners who live in ZEHs can make money by selling zero-emission electricity to others.

11.3 Energy Poverty in Smart City

The transition to smart cities can be an obstacle for poor households to access affordable energy and it can be a cause for households to fall into energy poverty. In smart cities towards carbon neutrality, electricity replaces fossil fuels and electricity prices are supposed to rise. As a result, the poor households without PVs need to pay more for energy and become poorer. Households with PVs and BEVs benefit greatly from the energy transition in smart cities, but poor households that cannot afford them get little from it. The poor households will lose access to affordable energy, which is now a part of Sustainable Development Goals (SDGs 7).

11.3.1 Energy Poverty and Fuel Poverty

"Energy poverty" concerns the accessibility of poor households to energy at an affordable price. The concept of "energy poverty" is derived from the concept of "fuel poverty". Fuel poverty refers to a situation in which poor households lack access to affordable fuels to heat their houses sufficiently in cold regions such as the United Kingdom and Scandinavia. In cold regions, people easily become ill if they do not have a warm living environment during winter.

Boardman (1991) is one of the most frequently mentioned primary studies on fuel poverty. Based on the traditional definition of fuel poverty, that is, "fuel poverty is the inability to afford adequate warmth because of the energy inefficiency of the home". Boardman, (1991, p. 201) proposed that "if a household's energy costs are above 10 percent of their income, they should be entitled to receive additional assistance". Fuel poverty differs from merely being poor, as low-income, energy-inefficient buildings, and high fuel prices together incur fuel poverty. The Warm Homes and Energy Conservation Act of 2000 in the United Kingdom was well-known as an earlier legislation that dealt with fuel poverty. Section 1 of the act stipulated, "a person is to be regarded as living 'in fuel poverty' if he is a member of a household living on a lower income in a home which cannot be kept warm at reasonable cost".

The UK Fuel Poverty Strategy of 2001 measured fuel poverty using the 10% standard that a fuel-poor household needs to spend more than 10% of its income on fuel and to heat its home to an adequate standard of warmth. According to Annual Fuel Poverty Statistics Report (DECC, 2013), which explained the 10% definition in detail, though "the definition is on heating the home, modelled fuel costs in the definition of fuel poverty also include spending on heating water, lights and appliance usage and cooking costs" (DECC, 2013, p. 5). In addition, it introduced a low-income, high-cost definition in addition to the 10% standard. The low-income, high-cost definition considers that a household is fuel poor where they need to pay fuel cost above average and "they would be left with a residual income below the official poverty line" (DECC, 2013, p. 8).

As research on fuel poverty has spread to the United States and other countries for various climates and temperatures, both winter heating and summer cooling are within the scope of fuel poverty research (Bouzarovski and Petrova, 2015, p. 33; Sánchez, Mavrogianni and González, 2017, pp. 344–345). In the literature, the term "energy poverty" is often used, while fuel poverty and energy poverty have been used interchangeably. For example, Harrison and Popke (2013, p. 249) used the term energy poverty to describe a situation in which "a household cannot afford to maintain the home's indoor temperature at a level that allows for a comfortable or healthy lifestyle, a condition also known as fuel poverty". Mohr (2018, p. 360) defined fuel poverty as "one term used to describe households' inability to afford energy for essential services such as heating, cooling, and lighting in the home". Li *et al.* (2014, pp. 479–480) analysed differences in how energy poverty and fuel poverty were used and found that:

> energy poverty addresses basic issues of energy access, whereas fuel poverty focuses on the issues of affordability and that fuel poverty mostly occurs in relatively wealthy countries with cold climates, whereas energy poverty occurs across all climates but mostly in poor countries.

Japan's major issue in energy poverty is not access to clean energy such as electricity but rather the affordability of adapting to new energy systems in a smart city. Access to affordable electricity is secured by the universal service system of the Electricity Business Act (§17(3) and §21). Even households in small villages on islands and in mountainous areas can access affordable electricity. From the perspective of energy poverty, the concern is whether poor households will be able to pay the cost of adapting to new energy systems in smart cities. Unless their houses are equipped with PVs and storage batteries, they will continue paying expensive electricity bills and their residual income will fall below the official poverty line owing to the large expenditure on electricity.

11.3.2 Energy Poverty Research in Japan

The concept of energy poverty or fuel poverty was introduced in Japan through fuel poverty policies introduced from abroad, particularly in the United Kingdom. The year 2017 was an exciting year for energy poverty research in Japan. Academic papers focusing on energy poverty were published in Japanese for the first time. Okushima (2017) analysed energy poverty by considering the rising energy vulnerability in 2013 caused by the 2011 Fukushima nuclear accident. Using indices such as energy cost, income, and house quality, he concluded that the energy poverty rate increased after 2000 and that 12.5% of elderly people were in energy poverty. He showed that climate policies such as FITs and carbon tax functioned as regressive tax rates and that those policies were unpopular among poor households. Thus, he discusses energy poverty in terms of energy justice.

Uezono (2017) examined energy poverty in Japan in a study that analysed German climate policy. He recognised that climate change mitigation policies can reduce energy poverty through energy saving but that poor households would not afford to engage in energy-saving behaviours prompted by government subsidies (Uezono, 2017, p. 63). Subsequently, Kumagai (2021, 2023) discussed how energy poverty in Europe was worsening because of increasing oil and gas prices. Kato (2020) analysed energy poverty from the perspective of policy consolidation. Kato (2022) reviewed research on energy poverty in Japan and studied an index for energy poverty measurement. Konno, Mori, and Iwata (2018) investigated fuel poverty in Hokkaido. Hokkaido, located in northern Japan, has very low temperatures in winter. They found that the fuel poverty rate was strongly correlated with the rate of houses built before 1980 and the rate of household income of less than 3 million JPY. There are few Japanese papers on energy poverty from a legal perspective (Kurokawa, 2022).

Although academic literature on energy poverty in Japan written in Japanese is not abundant, several articles have been written in English. A leading scholar in this field has been Okushima who authored several papers. The first study in English was even published earlier than his Japanese paper. Okushima (2016) examined fuel poverty by using the conventional 10% definition and showed that "energy poverty has worsened in Japan, especially for lower-income households and vulnerable households (mother–child and single-aged households) after the 2000s" (Okushima, 2016, pp. 562–563), and that energy prices became the main factor influencing energy poverty after the Fukushima nuclear accident. More precisely, he showed that "the proportion of households experiencing energy poverty in the lowest income decile group steadily increases from 34.6% in 2004 (the initial year) to 47.5% in 2013 (the latest year)". Okushima (2019) investigated energy service usage and revealed the importance of kerosene in northern Japan. He suggested that the energy transition from fossil fuels to renewables be promoted carefully with due consideration of its impact on the energy-poor because "there is a trade-off between climate change mitigation and energy poverty alleviation" (Okushima, 2019, p. 181). Chapman and Okushima (2019) investigated the relationship between energy poverty and low-carbon transition in Japan. They found that households experiencing energy poverty had a negative attitude towards the use of solar power driven by FITs and that they were less interested in low-carbon energy transition.

Tabata and Tsai (2020) study should be mentioned, which focused on Japan's hot and humid climate and the cooling expenditure. These authors concluded that "fuel poverty in summer is extremely low compared with that observed in winter" (Tabata and Tsai, 2020, p. 10). They also found that rising energy expenses due to carbon pricing were a significant factor contributing to the vulnerability of elderly households to fuel poverty. In the light of presented research, energy transition may pose challenges for poor households in some cases, potentially increasing electricity bills. Therefore, mitigating the negative impact of energy transition on poor households should be considered during the policymaking with respect to carbon neutrality.

11.3.3 Structure of Electricity Bill in Japan

In Japan, the electricity prices for residences were completely liberalised in 2016. However, most households have not shifted from the previously regulated electricity fee plan, and the former continues to prevail in electricity contracts with households. For households in poverty, the metered power plan B is usually applied. Electricity fees depend on the amount of electricity used. For example, considering the fee structure of the TEPCO, which has the largest share of the Japanese electricity market, the electricity bill consists of base fees, electricity charges, fuel cost adjustment fees, and renewable energy surcharges.

After the TEPCO electricity rate revision in June 2023, regulated electricity charges for household use increased in each category: 0–120 kWh, 30 JPY/kWh; 120–300 kWh, 36.6 JPY/kWh; and over 300 kWh, 40.69 JPY/kWh (TEPCO, 2023a). The price increased by approximately 11% owing to the soaring prices of fuel resources in international markets. As of February 2024, the fuel cost adjustment fee for low-voltage contracts is –9.56 JPY/kWh, reflecting a 3.5 JPY/kWh governmental subsidy called "electricity and gas price volatility relief measure" (TEPCO, 2024a).

11.4 Mobility in Smart City

In a smart city aiming to reach carbon neutrality, vehicles powered by fossil fuels should be replaced by electric vehicles such as BEVs and fuel cell vehicles (FCVs). In addition, people should use public transportation systems, such as subways, light rails, electric buses, and fuel cell buses, because they carry passengers with less energy consumption than passenger cars. The Japanese government has promoted the introduction of electric buses with subsidies. One-third of the bus price was covered by the subsidy if bus businesses bought electric buses in 2023 (Ministry of Land, Infrastructure, Transport and Tourism, MLIT, 2023b). However, because "compared to EVs, FCVs have the advantage of longer cruising ranges and shorter refuelling times" the government has also promoted FCVs and "supported the initiative to install hydrogen refuelling stations in advance of deploying FCVs" (MCREH, 2023, p. 42). Owing to government support, the cost of introducing non-fossil fuel buses has not been reflected in bus fares to date. Although transportation expenses are not an energy expense, increases would negatively impact the budgets of poor households that do not own cars and would cause a problem similar to energy poverty.

In addition to public transportation, passenger cars must shift to electric vehicles. Hydrogen and e-fuel can be alternatives to fossil fuels for internal combustion engine vehicles; however, BEVs are currently the most practical option. Presently, BEV owners enjoy cheaper energy costs than the owners of gasoline cars, if BEV owners charge their vehicles at home. The BEV charge cost is lower than that of gasoline fuel. For example, when comparing Nissan Leaf, which is a representative EV in Japan, with the Toyota Corolla Sport, which has the same body size as Leaf, the energy cost of Leaf is 60% of that of the Corolla Sport. According to their respective catalogue data, the Corolla Sport GX has a fuel economy of 18.3 km/l (WLTC mode), and the power consumption rate of Leaf X with a 40 kWh battery is 0.155 kWh/km (WLTC mode). In December 2023, the gasoline price was 175 JPY/l (ANRE, 2023a) and the electricity price was 36.6 JPY/kWh (TEPCO, 2023 b). To drive for 1,000 km, the Corolla Sport and Leaf cost 9,562 yen and 5,673 JPY, respectively. Therefore, those with BEVs and charging facilities at home can benefit from a BEV shift. However, BEVs cost more than gasoline cars for poor people who live in apartment complexes and cannot charge at home, because charging at a charging station costs twice as much as charging at home. It can be said that the shift to BEVs brings benefits to rich people but increases the energy expenditure of poor people. To mitigate this injustice, hybrid electric vehicles (HEVs) may be a good option for poor people because HEVs run for twice as long as gasoline vehicles with the same amount of gasoline. For example, in the case of the Corolla Sport GX, the fuel economy of the HEV is 30 km/l (WLTC mode), whereas that of a traditional gasoline vehicle is 18.3 km/l (WLTC mode).

The Japanese government has set a target for electrified vehicles (BEVs, FCVs, HEVs, and plug-in HEVs) to account for 100% of the sales of new passenger cars by 2035 (METI *et al.*, 2021, p. 78). Owing to governmental subsidies, purchasing a BEV itself does not currently incur additional payments in Japan. BEVs are also expected to be storage batteries for houses and to be managed by HEMSs. Vehicle-to-home (V2H) technology has been incorporated into Japan's energy transition policy. Specifically, V2H increases resilience to natural disasters (METI and MLIT, 2020).

If BEVs are charged with electricity generated on rooftops, the cost is almost zero, and the vehicle runs emitting no CO_2. Charging BEVs with electricity from electricity companies leads to CO_2 emissions because two-thirds of the electricity in Japan comes from thermal power plants (IEA, 2022). Given CO_2 emissions, "well-to-wheel" should be considered rather than "tank-to-wheel". The 2030 fuel efficiency standards are applying to BEVs and plug-in HEVs based on the well-to-wheel calculations (METI and MLIT, 2019).

11.5 Energy Poverty and Energy-Efficient Houses

Because heating accounts for approximately half of the energy consumption in the residential sector, thermal insulation and the use of highly efficient home appliances contribute to reducing energy consumption (MLIT, 2022, p. 13). Houses insulated from outside cold temperatures remain warm with less energy consumption. It was stated that "improved energy efficiency is the permanent non-reversible component of reducing energy poverty" (Boardman, 2010, p. 125). Helping poor households live in energy-efficient houses simultaneously contributes to reducing energy poverty and CO_2 emissions. Poor households usually do not have enough money to rent insulated energy-efficient houses even though they know that insulated houses are financially better off in the long run because of their lower heating costs. In this context, the fuel allowance in winter in cold Hokkaido can be recognised as part of the rent allowance.

The Building Energy Saving Act (2015) regulates the energy efficiency of buildings. Legislation requires every new building, including residences, to meet energy-efficiency standards after April 2025. The Strategic Energy Plan of 2021 projects that the energy-efficiency performance of houses and buildings that will be constructed after 2030 will reach the level of the ZEH standard (ANRE, 2021, p. 6). The Act on Promotion of Quality Assurance of Housing (1999) requires those selling or renting houses to disclose their energy efficiency to customers. In addition, the government has established insulation performance and energy-efficiency grading systems. Energy-efficient houses are too expensive for poor households. They typically rent inexpensive apartments that are insufficiently insulated.

11.5.1 PVs and Storage Batteries in Houses

If the PV capacity is sufficiently large to cover the electricity demand of the household and a storage battery is installed in the house, the household generally does not need to purchase electricity. If a household generates more electricity than it consumes, it can sell the surplus electricity to an electricity company. Therefore, households equipped with PVs and storage batteries do not fall into energy poverty.

In 2023, installing a PV on a newly built house rooftop costs 0.288 million JPY/kWh on average, such that installing a 3 kW capacity PV cost 0.864 million JPY. The same cost on the rooftop of an existing house was 0.834 million JPY according to ANRE (2023b, p. 37). The household that installed the 3 kW capacity PV received a subsidy from the Tokyo Metropolitan Government (TMG), while 0.45 million JPY was received in case of the new house, and 0.48 million yen in case of the existing house in 2023. The FIT scheme for PVs on house rooftops is a net-metering scheme. Most households with PV systems sell surplus electricity under the FIT scheme. In 2023, the FIT price for residential PVs under a 10 kW capacity was 16 JPY/kWh, although households bought electricity at over 30 JPY/kWh. Therefore, consuming electricity they generate, rather than selling it to electricity companies, reduces household electricity expenditure.

In Japan, PVs and storage batteries can be leased at reasonable prices. For example, the TEPCO group leases a PV system with a 3 kW capacity and an appropriate storage battery for 2,300 JPY a month after accepting government subsidies (TEPCO, 2024b). Under this lease contract, poor households with their own houses can use PVs and storage batteries. Consequently, they can save money in terms of energy consumption. However, it is difficult for poor people living in rental housing to install PVs and storage batteries because it is usually banned to remodel houses by the residential rental contracts in Japan.

11.5.2 ZEHs and LCCM

In 2018, the government of Japan formulated Strategic Energy Plan (fifth) and set a goal that more than half of newly built custom-built detached houses built by home builders would be ZEHs by 2020 and that all would be ZEHs by 2030 (ANRE, 2018, p. 41). The GX Realisation Basic Policy of 2023 requires new houses to meet the energy-efficiency performance standard by 2025 and achieve ZEH-level energy-efficiency performance by 2030 (Cabinet of Japan, 2023, p. 14). The TMG amended the Tokyo Environmental Protection Ordinance to require newly built houses to install PV systems. Major housing suppliers that supply over

20,000 m^2 housing units per year will be subject to this obligation from April 2025 (TMG, 2022).

In 2020, 62,560 ZEH custom-built houses were built upon customer request (MLIT, 2023c, p. 20). Among the custom-built houses built nationwide by leading major home builders, 56% were ZEHs. However, 24% of all custom houses built in 2020 were ZEHs, and 33.5% were ZEHs in 2022 (MLIT, 2023c; Sustainable Open Innovation Initiative, SII, 2023). In contrast, among all ready-built houses, ZEHs accounted for only 2.5% (3,246/129,351) in 2020 and 4.6% (6,587/144,321) in 2022 (SII, 2023, p. 15). Notably, the ZEH ratio for ready-built houses was extremely low compared with that of custom-built houses.

The government also promotes life-cycle carbon-minus (LCCM) houses, which generate more electricity than that consumed during their life-cycle, including during the construction, operation, and decommission–disposal periods. The national government encourages the construction of ZEHs and LCCM houses with subsidies of up to 1.4 million yen (MLIT, 2023a). Local governments have also prompted ZEHs to provide subsidies and tax reductions. For example, according to a TMG estimate (TMG, 2024), constructing a ZEH with a 4 kW capacity PV in Tokyo receives a 0.7 million JPY TMG subsidy and 0.8 million JPY national subsidy in addition to an annual 0.13 million JPY electricity bill reduction, despite the 1.78 million JPY spent for additional construction and installation costs. Most construction and installation costs are offset by government subsidies. Therefore, the monthly reduction in electricity payments belongs to households. Therefore, from a long-term perspective, Japan's current policy of encouraging ZEHs through subsidies can contribute to mitigating fuel poverty.

Residents in rental housing, including apartment complexes, can enjoy the benefits of ZEHs if the owners of rental housing provide ZEHs to their residents. The national government encourages the owners of apartment complexes to build ZEH. ZEH mansion (ZEH-M), nearly ZEH-M, ZEH-M-ready, and ZEH-M-oriented apartment buildings are also encouraged with subsidies. These policies contribute significantly to the prevention of energy poverty because households in poverty tend to live in apartments.

A ZEH stores excess electricity in storage batteries during the day and discharges the necessary electricity from the storage batteries at night. As the retail electricity price increases, it becomes cheaper to generate and consume electricity at home. In 2023, the second-stage standard electricity fee of TEPCO was 36.6 JPY/kWh, whereas the FIT price for rooftop PVs was 16 JPY/kWh. Therefore, households can save more money by storing excess electricity in storage batteries than by exporting it during the day under the FIT scheme and buying electricity at night. In addition, ZEHs eliminate the renewable energy charge of the FIT scheme, as households purchase little electricity from an electricity company.

A serious problem related to energy poverty is that poor households cannot afford to live in new apartments that are insulated and equipped with PVs. The rent of these high-performance rental houses is usually expensive and beyond their budgets. It is true that energy poverty differs from poverty alone, but it is difficult for poor households to access energy-efficient houses with PVs and storage batteries.

11.6 HEMSs and Smart Grid

Generally, electricity generated from wind and solar power is unstable because it depends on weather conditions. Sunlight is easily shaded by clouds, and the wind easily changes its speed. The amount of electricity entering an electricity grid must always equal the amount of electricity consumed, to maintain voltage and frequency stability. Imbalances result in blackouts, which threaten the security of society. When the electricity supply exceeds the demand, grid companies require power generators to reduce the volume of electricity fed, to balance the supply and demand. Electricity from renewable energy sources is expected to be stored during periods of oversupply and discharged during peak demand.

Smart grids have been developed to balance the supply and demand of electricity. The smart grid concept encompasses power generators, transmission and distribution systems, and power consumers (New Energy and Industrial Technology Development Organization of Japan, 2014, p. 533). Distributed electricity resources (DERs), such as small PVs, small wind farms, small gas-fired power generators, and storage batteries, are controlled by a smart grid equipped with information and communication technology and artificial intelligence. It is estimated that smart grids will control electricity generation and consumption via HEMSs, for example by raising or lowering the temperature of air conditioners or maintaining hot water in tanks. The adjustment capacities created by the respective houses are small; however, their aggregation had sufficient potential to balance the gap. The legal status of the electricity supply and demand aggregator was stipulated for the first time by the 2020 amendment to the Electricity Business Act of Japan. Aggregators aggregate DERs and small demands, and they work as adjustment capacity and virtual power plants.

In Japan, households that become part of smart grids enjoy low electricity expenditures because they generate electricity for their own consumption. They can make reasonable electricity price contracts such as off-peak discount plans because they contribute to a peak shift in electricity demand. They use the electricity generated by PVs during the day and the electricity stored in the home storage batteries at night. For households with PVs but no storage batteries, TEPCO sells a surplus electricity deposit service.

Households left behind in the energy transition continue to pay expensive electricity bills. People living in poverty cannot afford to own houses in urban areas. They typically rent houses and live in apartment complexes. The Japanese government encourages zero-energy apartment complexes called ZEH-Ms. However, only 2.1% of new apartment complexes were ZEH-Ms in 2021, whereas approximately half of the residents of ZEH-Ms answered that their electricity bills decreased after they moved to ZEH-Ms (SII, 2022, p. 16, p. 218). As apartment renters build

ZEH-Ms to set higher rents, it is not economically easy for poor families to move into ZEH-Ms. Therefore, they fall into energy poverty when electricity prices increase further.

11.7 Conclusion

To reduce CO_2 emissions in Japanese cities, energy consumption should shift from fossil fuels to electricity consumption, and consumed electricity should be generated from non-emission energy sources. Renewable energy growth can be realised using grid management to balance electricity supply and demand. Residences equipped with PVs and storage batteries contribute to renewable energy penetration and help balance electricity supply and demand in cities. BEVs can act as a complementary solution to home storage batteries. Households with PVs will pay smaller energy bills because they consume electricity they generate on their rooftops.

Despite the potential advantages, there are certain challenges to address. Energy transition in smart cities appears to mitigate energy poverty problems. However, households in poverty usually live in low-rent rental houses that are not equipped with PVs and that are not well insulated. Therefore, most households in poverty could be left behind in the energy transition and will not enjoy its benefits unless the government formulates an appropriate policy to help households in poverty. Without such a policy, energy poverty would become an obstacle to the energy transition towards a low-carbon society.

Note

1 FY2021 in Japan covered the period from 1 April 2021 to 31 March 2022.

References

Act on Promotion of Quality Assurance of Housing (1999) No 81. In Japanese: 住宅の品質確保の促進等に関する法律 [juhtaku-no hinshitukakuho-nosuishintou-ni kansuru houritsu].

ANRE (2018) "5th Strategic Energy Plan". In Japanese: エネルギー基本計画(第5次) [enerugī kihon keikaku (dai-go-ji)]. Available at: www.enecho.meti.go.jp/en/category/others/basic_plan/5th/pdf/strategic_energy_plan.pdf (Accessed: 27 March 2024).

ANRE (2021) "6th Strategic Energy Plan". In Japanese: エネルギー基本計画(第6次) [enerugī kihon keikaku (dai-roku-ji)]. Available at: www.enecho.meti.go.jp/category/others/basic_plan/pdf/strategic_energy_plan.pdf (Accessed: 27 March 2024).

ANRE (2023a) "Petroleum Products Price Research Results". Available at: www.enecho.meti.go.jp/statistics/petroleum_and_lpgas/pl007/results.html (Accessed: 27 March 2024).

ANRE (2023b) "Presentation Material: Solar Photovoltaic". Available at www.meti.go.jp/shingikai/santeii/pdf/091_01_00.pdf (Accessed: 27 March 2024).

ANRE (2023c) "News Release: FY2022 Comprehensive Energy Statistics (Preliminary Figures)". Available at: www.enecho.meti.go.jp/statistics/total_energy/pdf/gaiyou2022fysoku.pdf (Accessed: 27 March 2024).

Boardman, B. (1991) *Fuel Poverty: From Cold Homes to Affordable Warmth*. London: Belhaven Press.

Boardman, B. (2010) *Fixing Fuel Poverty: Challenges and Solutions*. London: Routledge.

Bouzarovski, S. and Petrova, S. (2015) "A Global Perspective on Domestic Energy Deprivation: Overcoming the Energy Poverty-Fuel Poverty Binary", *Energy Research & Social Science*, 10, pp. 31–40.

Building Energy Saving Act – Act on the Improvement of Energy Consumption Performance of Buildings Control (2015) No. 53. In Japanese: 建築物のエネルギー消費性能の向上等に関する法律 [Kenchiku-butsu no enerugī shōhi seinō no kōjō-tō ni kansuru hōritsu].

Cabinet of Japan (2023) "The Basic Policy for the Realization of GX: A Roadmap for the Next 10 Years". Available at: www.cas.go.jp/jp/seisaku/gx_jikkou_kaigi/pdf/kihon_en.pdf (Accessed: 27 March 2024).

Chapman A. and Okushima S. (2019) "Engendering an Inclusive Low-Carbon Energy Transition in Japan: Considering the Perspectives and Awareness of the Energy poor", *Energy Policy*, 135, p. 111017.

Department of Energy and Climate Change (DECC) of the United Kingdom (2013) "Annual Report on Fuel Poverty Statistics 2013". Available at: https://assets.publishing.service.gov.uk/media/5a750ccde5274a3cb2869406/Fuel_Poverty_Report_2013_FINALv2.pdf (Accessed: 27 March 2024).

Eco-Town Act – City Planning Act for Promotion of Low Carbon Cities (2012) No. 84. In Japanese: 都市の低炭素化の促進に関する法律 [toshi-no teitansoka-no sokushin-ni-kansuru houritsu].

Electricity Business Act (1964) No. 170. In Japanese: 電気事業法 [denki jigyō-hō].

GX Promotion Act – Act on Promotion of Smooth Transition to a Decarbonized Growth-Oriented Economic Structure (2023) No. 32. In Japanese: 脱炭素成長型経済構造への円滑な移行の推進に関する法律 [datsu tanso seichō-gata keizai kōzō e no enkatsuna ikō no suishin ni kansuru hōritsu].

Harrison, C. and Popke, J. (2013) "'Because You Got to Have Heat': The Networked Assemblage of Energy Poverty in Eastern North Carolina", in K. Zimmer (ed.), *The New Geographies of Energy: Assessment and Analysis of Critical Landscapes*. London: Routledge, pp. 248–260.

International Energy Agency (IEA) (2022) "Electricity Generation Mix, Japan, 2022". Available at: www.iea.org/countries/japan/energy-mix (Accessed: 27 March 2024).

Kato, R. (2020) "環境・経済・福祉の統合に向けて ： エコロジー的近代化からエネルギー貧困まで" [Toward the Integration of Environment, Economy, and Welfare: Rethinking Ecological Modernization and Energy Poverty], *経済科学* [*The Economic Science*], 67(3), pp. 29–39.

Kato, R. (2022) "エネルギー貧困の測定方法の検討" [Study on Measurement Method of Energy Poverty], *金沢学院大学紀要* [*Journal of Kanazawa Gakuin University*], 20, pp. 70–78.

Konno, Y., Mori, T. and Iwata, Y. (2018) "北海道におけるFuel Povertyの実態に関する研究" [Research on Actual Condition of Fuel Poverty in Hokkaido], *日本建築学会環境系論文集* [*Architectural Institute of Japan's Journal of Environmental Engineering*], 83, pp. 729–736.

Kumagai, T. (2021) "欧州市民を襲う異常価格: エネルギー貧困が現実問題に" [Soring Energy Prices Hit Europeans: Energy Poverty Becomes a Real Problem], *週刊東洋経済* [*Weekly Oriental Economist*]. Available at: https://toyokeizai.net/articles/-/575928 (Accessed: 27 March 2024).

Kumagai, T. (2023) "欧州が陥る「エネルギー貧困」の現実熊谷 徹" [Reality of energy poverty in EU], *Voice*, 552, pp. 72–79.

Kurokawa, S. (2022) "SDGs", エネルギー正義、そして再生可能エネルギーの拡大 [SDGs, Energy Justice, and Renewable Energy Penetration], 環境法研究 [*Environmental Law Studies*], 47, pp. 42–56.

Li, K. *et al.* (2014) "Energy Poor or Fuel Poor: What Are the Differences?", *Energy Policy*, 68, pp. 476–481.

METI (2023) "Comment on Procurement Price after 2023 by Calculation Committee for Procurement Price Document". Available at: www.meti.go.jp/shingikai/enecho/shoene_shinene/suiso_seisaku/pdf/20230606_5.pdf (Accessed: 27 March 2024).

METI and MLIT (2019) "「乗用車の新たな燃費基準に関する報告書」の公表" [Publication of the 'Report on New Fuel Efficiency Standards for Vehicles']. Available at: www.mlit.go.jp/common/001295001.pdf (Accessed: 27 March 2024).

METI and MLIT (2020) "災害時における電動車の活用促進マニュアル" [Manual for Promoting the Use of Electric Vehicles in the Event of a Disaster]. Available at: https://warp.da.ndl.go.jp/info:ndljp/pid/13120268/www.meti.go.jp/press/2020/07/20200710006/20200710006-1.pdf (Accessed: 27 March 2024).

METI *et al.* (2021) "Green Growth Strategy through Achieving Carbon Neutrality in 2050". Available at: www.meti.go.jp/english/policy/energy_environment/global_warming/ggs2050/pdf/ggs_full_en1013.pdf (Accessed: 27 March 2024).

Ministerial Council on Renewable Energy, Hydrogen and Related Issues of Japan (MCREH) (2023) "Basic Hydrogen Strategy". Available at: www.meti.go.jp/shingikai/enecho/shoene_shinene/suiso_seisaku/pdf/20230606_5.pdf (Accessed: 27 March 2024).

MLIT (2022) "Summary of the White Paper on Land, Infrastructure, Transport and Tourism in Japan 2022". Available at: www.mlit.go.jp/en/statistics/white-paper-mlit-index.html (Accessed: 27 March 2024).

MLIT (2023a) "令和5年度LCCM住宅整備推進事業" [FY2023 LCCM Housing Development Promotion Project]. Available at: www.mlit.go.jp/jutakukentiku/house/shienjigyo_r5-04.html (Accessed: 27 March 2024).

MLIT (2023b) "自動車環境総合改善対策費補助金（事業用自動車における電動車の集中的導入支援）事業（令和４年度補正)" [FY2022 Supplementary Budget: Vehicle Environmental Improvement Subsidy Project (Intensive Support for the Introduction of Electrified Vehicles in Commercial Fleets)]. Available at: www.mlit.go.jp/jidosha/jidosha_tk10_000040.html (Accessed: 27 March 2024).

MLIT (2023c) "Summary of the White Paper 2022 on Land, Infrastructure, Transport and Tourism in Japan". Available at: www.mlit.go.jp/en/statistics/content/001579732.pdf (Accessed: 27 March 2024).

MoE (2023) "令和３年度 家庭部門のCO_2排出実態統計調査結果の概要（確報値）令和５年３月)" [Outline of Research Report on CO_2 Emission from Households Sector in FY2021]. Available at: www.env.go.jp/content/000122573.pdf (Accessed: 27 March 2024).

MoE and NIES (2023). "2021年度（令和３年度）の温室効果ガス排出・吸収量（確報値1） について" [The Confirmed Greenhouse Gas Emissions and Absorption Data for Fiscal Year 2021]. Available at: www.env.go.jp/content/000128750.pdf (Accessed: 27 March 2024).

Mohr, T.M. (2018) "Fuel Poverty in the US: Evidence Using the 2009 Residential Energy Consumption Survey", *Energy Economics*, 74, pp. 360–369.

New Energy and Industrial Technology Development Organization (2014) "NEDO再生可能エネルギー技術白書９スマートグリッドの技術の現状とロードマップ" [White

Paper on Renewable Energy Technology, 2nd edition]. Available at: www.nedo.go.jp/library/ne_hakusyo_index.html (Accessed: 27 March 2024).

Okushima, S. (2016) "Measuring Energy Poverty in Japan, 2004–2013", *Energy Policy*, 98, pp. 557–564.

Okushima, S. (2017) "「エネルギー貧困」・「エネルギー脆弱性」・「エネルギー正義」：日本における現状と課題" [Energy Poverty, Energy Vulnerability and Energy Justice: Present Situation and Future Issue in Japan], *科学* [*Science*], 87(11), pp. 1019–1027.

Okushima, S. (2019) "Understanding Regional Energy Poverty in Japan: A Direct Measurement Approach", *Energy and Buildings*, 193, pp. 174–184.

Sánchez, C.S.G., Mavrogianni, A. and González, F.J.N. (2017) "On the Minimal Thermal Habitability Conditions in Low-Income Dwellings in Spain for a New Definition of Fuel Poverty", *Building and Environment*, 114, pp. 344–356.

SII (2022) "ネットゼロ・エネルギー・ハウス実証事業調査発表会2022のご案内" [Net-Zero-Energy-House Demonstration Project Research Presentation 2022]. Available at: https://sii.or.jp/zeh/conference_2022.html (Accessed: 27 March 2024).

SII (2023) "ネット・ゼロ・エネルギー・ハウス実証事業調査発表会 2023" [Net-Zero-Energy-House Demonstration Project Research Presentation 2023]. Available at: https://sii.or.jp/meti_zeh05/uploads/ZEH_conference_2023.pdf (Accessed: 27 March 2024).

Sokołowski, M.M. (2022) *Energy Transition of the Electricity Sectors in the European Union and Japan: Regulatory Models and Legislative Solutions*. Cham: Palgrave Macmillan.

Sokołowski, M.M. and Kurokawa, S. (2022), "Energy Justice in Japan's Energy Transition: Pillars of Just 2050 Carbon Neutrality", *Journal of World Energy Law & Business*, 15(3), pp. 183–192.

Tabata, T. and Tsai, P. (2020) "Fuel Poverty in Summer: An Empirical Analysis Using Microdata for Japan", *Science of the Total Environment*, 703, p. 135038.

TEPCO (2023a) "電気料金単価表" [Electricity Rate Table]. Available at: www.tepco.co.jp/ep/private/plan/pdf/teiatsu_minaoshi_pdf3.pdf (Accessed: 27 March 2024).

TEPCO (2023b) "料金単価表-電灯" [Electricity Rate – Price Chart]. Available at: www.tepco.co.jp/ep/private/plan/chargelist01.html (Accessed: 27 March 2024).

TEPCO (2024a) "燃料費調整のお知らせ（2024年2月分）" [Notification of fuel cost adjustment fee (February 2024)]. Available at www4.tepco.co.jp/ep/private/fuelcost2/backnumber/2402.html (Accessed: 27 March 2024).

TEPCO (2024b) TEPCO Group's Flat-Rate Equipment Usage Service. Available at www.tepco-ht.co.jp/enekari/lp/zero/?utm_source=yahoo&utm_medium=cpc&utm_content=1251000_13623002317_124101716556_648030264690_kwd-1187872484399&utm_campaign=01NQ_SCH_%28%E5%A4%AA%E9%99%BD%E5%85%89%E3%83%BB%E8%93%84%E9%9B%BB%E6%B1%A0%29-%E6%9D%B1%E4%BA%AC%E3%83%BB%E7%A5%9E%E5%A5%88%E5%B7%9D%E4%BB%A5%E5%A4%96&utm_term=e_%E3%82%A8%E3%83%8D%E3%82%AB%E3%83%AA%20tepco&yclid=YSS.1001196054.EAIaIQobChMIoN6R2ZmWhQMVgwZ7Bx0hmwVWEAAYASAAEgKbefD_BwE (Accessed: 27 March 2024).

TMG (2022) "Tokyo Solar Power". Available at: www.kankyo.metro.tokyo.lg.jp/en/climate/solar_portal/index.html (Accessed: 27 March 2024).

TMG (2024) "「東京ゼロエミ住宅」とは？" [What Is Tokyo Zero Emission House?]. Available at: www.kankyo.metro.tokyo.lg.jp/climate/home/tokyo_zeroemission_house/gaiyou.html (Accessed: 27 March 2024).

Uezono, M. (2017) "地球温暖化対策とエネルギー貧困対策の政策統合ー ドイツの省エネ診断制度を事例に" [Policy Integration Between Global Warming Measures and Energy Poverty Measures: German Energy Saving Diagnosis System], *経済科学論集* [*Journal of Economic Science*], 43, pp. 63–85.

United Nations Department of Economic and Social Affairs (2018) "World Urbanization Prospects the 2018 Revision". Available at https://population.un.org/wup/Publications/Files/WUP2018-Report.pdf (Accessed: 27 March 2024).

12 Can Toyota's Woven City Be Considered a Smart City?

Hiroshi Ito and Kazuhiko Kato

12.1 Introduction

The term "smart city" was first coined in the 1980s (Holguín Rengifo, Herrera Vargas and Valencia-Arias, 2023). Amsterdam's Digital City, established in 1994, is often cited as the world's first smart city (Dameri, 2017; Matsumoto, Esaka and Izumiyama, 2023). A smart city can be defined as a digitally interconnected and sustainable urban centre that serves as a crucible for user-driven innovation ecosystems (Scuotto, Ferrais and Bresciani, 2015). It comprises multi-dimensional factors intertwined with technology, services, and residents' discerning perspectives (Caputo, Buhnova and Walletzky, 2018). Central to the smart city paradigm is an emphasis on information and communication technology (ICT) and the integration of human and social capital as pivotal constituents (Singh and Singla, 2020). Moreover, smart cities leverage ICT to enhance the efficiency of conventional networks and services to better meet the needs of residents and businesses (Kummitha, 2019). This efficiency is achieved through data collection and utilisation (Spicer and Zwick, 2021) along with the optimisation of city operations, including energy consumption, waste management, water usage, citizen mobility, and crime prevention (Kiyatama, 2021).

Within the smart city framework, the concept of a corporate smart city is notable. Specifically, a corporate smart city is characterised by its reliance on corporate giants, such as Cisco and IBM, for technologies and services. Such cities create an entrepreneurial urban environment featuring experimental living labs, wherein technologies generate data, enabling enterprises to venture into novel prospects (Kummitha, 2019). This paradigm is mainly driven by private interests aspiring to gain control over public assets and services by offering technological solutions for urban challenges, such as congestion, emergency response, and the efficient delivery of utilities and service provision (Cardullo and Kitchin, 2019).

Prominent examples of corporate smart cities include Masdar City in Abu Dhabi, United Arab Emirates; Songdo in Incheon, South Korea; the Fujisawa Sustainable Smart Town (FSST) in Kanagawa, Japan; and Quayside in Toronto, Canada. Masdar City has been operated by key players such as the Abu Dhabi Future Energy Company and Siemens. Meanwhile, companies such as Gale International, IBM, and Cisco have collaborated to establish and develop Songdo. FSST benefits from

DOI: 10.4324/9781003471417-12

a strategic affiliation with Panasonic. Further, the Quayside project, spearheaded by Google's Sidewalk Labs (which was discontinued in 2020), exemplified the dynamic interplay between governments and corporations in the pursuit of urban digital transformation.

In alignment with this prevailing global trend, Woven by Toyota (WbyT), a subsidiary of Toyota Motor Corporation, has been engaged in Woven City, a unique urban project taking shape on the grounds of Toyota's former East Fuji Plant in the city of Susono in Shizuoka prefecture. Woven City presents an interesting case study: while it is physically located within Susono, it is being built entirely from the ground up as "a green field" development, with residents recruited exclusively from outside. It is also the first endeavour of this kind to be undertaken by a car-mobility company. However, comprehensive and publicly available information about Woven City has been scarce hitherto. In light of this, the current chapter delves into the features of Woven City, by reviewing relevant documents and interviewing stakeholders, with the aim of determining whether it can be considered a corporate smart city.

The structure of this study unfolds as follows. In the following section, the previous examples of corporate smart cities developed from scratch are explained (i.e. Masdar City, Songdo, Fujisawa, and Quayside). These examples help in identifying how Woven City may be different from other corporate smart cities. After the methodology and results sections, the findings from the document analysis are compared with those from interview studies in the discussion section to further elaborate on what Woven City would be like and specifically on whether it can be considered a corporate smart city. This is followed by the conclusion.

12.2 Methodology

In the present research, we adopted a case study approach. The rationale for a single case study emerges in an extreme or unusual case that deviates from theoretical norms while being of general public interest, and with underlying nationally important issues (Yin, 2018). Woven City is novel in the sense that it is arguably the first smart city to be built by a car-mobility company that not only manufactures cars but focuses on the broader aspects of mobility, including the services and technologies that enable the movement of people and goods. It is also an important project for both Toyota and the Japanese economy (Kitayama, 2021).

12.2.1 Methods

For data collection, we employed document analysis and semi-structured interviews. For document analysis, non-technical literature, such as reports and internal correspondence, can be used as empirical data for conducting case studies that focus on a singular phenomenon, event, organisation, or programme (Bowen, 2009). Furthermore, document analysis assumes a pivotal role in formulating interview questions (Bowen, 2009). Additionally, document analysis can mitigate certain ethical considerations linked to alternative qualitative methodologies (Morgan, 2022).

Table 12.1 Interview respondents

Respondents	*Affiliation*
T-1	Toyota Group
T-2	Toyota Group
T-3	Toyota Group
T-4	Toyota Group
T-5	Toyota Group
F-1	FSST
G-1	Susono City Government
A-1	University of Shizuoka

We also conducted semi-structured interviews. The rationale for using multiple sources of evidence in case study research relates to the motive of conducting an in-depth study of a phenomenon in its real-world context (Yin, 2018). Interviews can be an important source of case study evidence, providing explanations of key events as well as insights reflecting participants' relative perspectives (Yin, 2018).

Interviewees were selected using convenience sampling starting with Toyota employees studying at the Nagoya University of Commerce and Business. In total, six Toyota personnel were involved. We also conducted interviews with one FSST employee, two city government officials, two university professors, and four civil society organisation personnel. However, we only used the responses of the five Toyota employees, one FSST staff, one city government official, and one university professor due to their relevance to this study. In selecting participants, we prioritised individuals with expertise and perspectives directly pertinent to the central research question: "Can Toyota's Woven City be considered a smart city?" While additional interviews were conducted to ensure a comprehensive understanding of the context, only responses that directly illuminated the inquiry were included for detailed analysis in this study. In total, the responses of eight participants were analysed (see Table 12.1).

Ethical approval was obtained from the Research Ethics Review Board of the Nagoya University of Commerce and Business in May 2023. The interviews took place between June and November 2023. The purpose of the study was explained before the interviews, and oral consent for participation in the study was obtained. The interview questions included: "Where does the name 'Woven' come from?", "How was the concept of Woven City born?", and "Is Woven a corporate smart city?". The average duration of the interviews was 60 minutes.

12.2.2 Data Analysis

For data analysis, we chose a concise and streamlined approach, with the primary focus on our central research question: "What would Woven City be – in particular, would it be categorised as a smart city?". It should be noted that owing to limitations in space and methodology, we refrained from employing more intricate data analysis methods, such as thematic or content analysis. Instead, we adopted a more direct and linear narrative approach that closely aligns with the sequential nature of the questions posed to our interviewees and their responses.

12.3 Examples of Corporate Smart Cities

The dawn of the 21st century has witnessed an unprecedented convergence of urban planning and digital technology, giving rise to the concept of corporate smart cities. Projects such as Masdar City, Songdo , FSST, and Quayside encapsulate this convergence, each representing a unique experiment in integrating sustainability, renewable energy, and climate-conscious urban living with advanced technology. These urban projects were thus selected in this study to explore the multifaceted nature of smart cities.

12.3.1 Masdar City

With an extensive area spanning 6 km^2, Masdar City was constructed by the Abu Dhabi Future Energy Company, also known as Masdar (Janajreh, Su and Alan, 2013). In 2006, the Abu Dhabi government allocated 22 billion USD to construct Masdar as a carbon-neutral, zero-waste city (Griffiths and Sovacool, 2020). The word *masdar*, meaning "the source" in Arabic,[1] underlines the city's aspiration to diminish reliance on conventional oil and gas and strive instead for more sustainable energy and economic systems by integrating renewable energy technologies (Griffiths and Sovacool, 2020). To realise these goals, a comprehensive infrastructural framework was devised, encompassing diverse elements, such as rooftop photovoltaic panels, concentrated solar thermal collectors, geothermal sources, and a waste-to-energy facility (Sgouridis and Kennedy, 2010). Masdar City is committed to excluding fossil fuel-powered vehicles from its confines, opting instead for an innovative "personal rapid transit" (PRT) system, supplemented by cycling and pedestrian networks (Randeree and Ahmed, 2019).

However, the project deviated from its original trajectory in multiple ways. Initial plans to construct expansive solar panels were hampered by local dust storms, leading to a substantial 40% reduction in solar power output. Consequently, Masdar City had to supplement its energy by procuring power from external sources (Cugurullo, 2018). Even water desalination, envisioned as solar-powered, necessitated a deviation from the original plan owing to practical constraints, with energy-intensive gas-powered desalination emerging as the solution (Randeree and Ahmed, 2019). Another notable change was in the flagship transportation strategy. The automated electric PRT system, initially perceived as a central component of the car-free vision, underwent significant setbacks and ended up serving only a minor portion of the city owing to its high costs. The project subsequently shifted towards prioritising electric vehicles (Cugurullo, 2018).

Recalibration of the environmental objectives led to a heightened emphasis on Masdar City's role as a "living lab" or a "testbed" for innovative environmental technologies, smart grids, and public and private sector partnerships (Randeree and Ahmed, 2019). Companies like Siemens have actively engaged with the city's mission, exemplified by Siemens' installation of its Middle East headquarters within Masdar City (Cugurullo, 2018). Ownership of Masdar City was substantially restructured in the late 2010s, with ownership transferred to the Mubadala Investment Company, a corporate entity that emerged in 2017 under Abu Dhabi's auspices after the merger of the Mubadala Development Company and the International Petroleum Company in 2016 (Griffiths and Sovacool, 2020). Currently, an incremental approach has been adopted in Masdar City, focusing on establishing a pioneering urban model for form, structure, and mobility (Griffiths and Sovacool, 2020). As part of this approach, an electric vehicle car-sharing initiative, in partnership with Ekar, offers Tesla vehicles for rent, reflecting the city's adaptability and commitment to sustainability (Griffiths and Sovacool, 2020).

Significant shifts in Masdar City's master plan underscore its evolving nature. Originally conceived as a zero-carbon entity, the objective was revised to pursue carbon neutrality instead. Criticisms have included the failure to achieve democratic ideals and environmental milestones, particularly concerning energy, water, and ecological impact management (Cugurullo, 2018). Nevertheless, Masdar City's inception marked a bold stride towards sustainable urban design, evident in the substantial investment made by the Abu Dhabi government. Despite deviations from its original blueprint, Masdar City continues to carve a distinctive path towards a sustainable future.

12.3.2 Songdo

Songdo International Business District is a ground-breaking smart city venture (Kuecker and Hartley, 2019). Songdo is described as a "living lab for the 21st century" modelled on New York City's Central Park (Strickland, 2011). It was built from scratch on 6 km^2 of reclaimed land along the waterfront in south Incheon, 65 km west of Seoul, South Korea. Songdo is an international free economic zone that allows foreign land ownership and tax benefits, enticing global businesses in ICT, sustainable energy, and biotechnology sectors (Kshetri, Alcantara and Park, 2014; Selinger and Kim, 2015).

The private sector provided much of the capacity for designing, building, and operating the project. Major companies involved included Gale International (US-based real estate), Posco Engineering & Construction (Korean construction *chaebol* – a large family-owned business conglomerate), IBM, Cisco Systems, and Korea Telecom Corporation (Angelidou, 2014). Notably, Gale and Posco have 70% and 30% stakes, respectively. Posco carried out land reclamation, developed the infrastructure, and relocated its corporate headquarters from Seoul to Songdo. Cisco Systems played the lead role in developing Songdo's ICT hardware. Infrastructure development and technology integration by these companies contributed to Songdo's status as one of the world's largest private real estate developments (Segel *et al.*, 2012). Residents can now control some functions of their homes remotely and interact through video from anywhere in the world using Cisco's telepresence system (Angelidou, 2014).

Songdo aligned with South Korea's national ICT visions such as "e-Korea Vision" (2002) and "Broadband IT Korea Vision for a knowledge-based economy" (2003) (Mullins and Shwayri, 2016). The construction of Songdo began in 2003 (Benedikt, 2016); as a compact urban model, it integrates diverse aspects of life, business, education, and culture within a well-designed space (Selinger and Kim, 2015).

In 2006, the Ministry of Information and Communication updated to the Ubiquitous-IT839 or U-IT-839 strategy, establishing the U-Korea General Plan (Mullins and Shwayri, 2016). The IT-839 plan refers to a space without any constraints in connecting people, technologies, and objects through free information exchange (Mullins and Shwayri, 2016). U-IT-839 was designed to reflect a vision of Korea as a knowledge-based economy through developing three new growth areas: convergence of ICT services through broadband, infrastructure network changes, and new hardware-related business changes (Mullins and Shwayri, 2016). Although the first phase of Songdo began in August 2009, its construction proceeded slowly, with delays attributed to budgetary constraints, stakeholder resistance, and challenges in attracting foreign investments (Angelidou, 2014). In essence, Songdo emerged as a transformative smart city project with private sector involvement and technological integration. However, its complexities and challenges highlight the dynamic nature of smart urban development.

12.3.3 Fujisawa Sustainable Smart Town (FSST)

The FSST is a pioneering project initiated by Panasonic in November 2010 and officially inaugurated in November 2014, in Fujisawa, Kanagawa prefecture, in the western Kanto region of Japan. Covering 0.19 km^2, FSST comprises approximately 1,000 households (600 houses and 400 apartments) and various commercial, wellness, welfare, and educational facilities. The project's core objective is to establish an innovative urban lifestyle based on cutting-edge energy, security, mobility, wellness, and community engagement solutions. FSST currently accommodates over 2,000 residents, living in smart homes equipped with solar panels, storage batteries, and a home energy-management system to optimise power consumption (FSST Management, 2023).

In the last few decades, most Japanese electronic appliance companies, including Panasonic, have faced hardships. According to the Ministry of Economy, Trade and Industry (METI), around 40% decrease in the output of electric goods was recorded over the 15 years from 2000 to 2015 (Japan Electrical Manufacturers' Association: JEMA, 2016). Although Panasonic recorded a sales peak of 9.1 trillion JPY (800 million USD) in 2007 (Panasonic, 2007), the figure had fallen to 7.6 trillion JPY (670 million USD) by 2016 (Panasonic, 2016).

Panasonic's motivation for building FSST is similar to that of IBM for assisting in building of smart cities worldwide. In the early 2000s, when the electronics industry showed a decline, Panasonic shifted its focus towards innovative town management systems even though it had never previously engaged in a town-development project. In September 2007, Panasonic, in collaboration with

PanaHome and the local government of Fujisawa, proposed the creation of the Fujisawa Eco-Town (Sakurai and Kokuryo, 2018). In March 2008, three factories of the Matsushita Electric Industrial Group (now Panasonic) located in the city of Fujisawa were closed (Sakurai and Kokuryo, 2018). The closure prompted the redevelopment of the site in Fujisawa. The city government asked Panasonic to redevelop the neighbouring areas. Panasonic established a project team to redesign the factory site for redevelopment to create an environmentally and socially sustainable town – FSST (Sakurai and Kokuryo, 2018).

The project's scope has expanded rapidly, necessitating the involvement of housing, real estate, finance, and consulting companies (Tokoro, 2018). A consortium of 18 organisations has been formed (as of 2023), with a focus on infrastructure provision and business-related services (FSST Management, 2023). FSST was not only a residential experiment but also a collaborative testing ground. It led to the commercialisation of various technologies, including sleep-monitoring sensors that adjust smart air-conditioner settings based on detected body movements (Hornyak, 2022). Residents access services through a unified online platform, connected to a "smart energy gateway" that facilitates energy consumption monitoring, sales notifications, and facility bookings. If they observe any problems or have any suggestions to address these problems, they are encouraged to report them to the FSST Management Team, either through the portal site or town meetings, for potential improvement (Tokoro, 2018).

The overarching goals of FSST include a goal of reducing CO_2 emissions by 70% and water consumption by 30% compared with the 1990 levels, along with achieving 30% electricity generation from renewable sources (compared with normal detached properties) (Barrett, DeWit and Yarime, 2021). The Japanese government financially supported FSST, contributing to the construction of low-emission housing and energy-related infrastructure. Half the cost of building low-emission housing and offices was financed by the Ministry of Land, Infrastructure and Tourism, and METI provided matching funds amounting to 9.4 billion JPY (86 million USD) (Barrett, DeWit and Yarime, 2021). The company envisaged promoting collective actions based on extensive information sharing in the town (e.g. reducing energy consumption, securing safety of the town, car sharing) by monitoring residents' activities and analysing collected data (Sakurai and Kokuryo, 2018). As a trailblazing initiative, FSST demonstrates the potential of smart urban development in addressing sustainability challenges and transforming lifestyles.

12.3.4 Quayside, Waterfront Toronto

Although Quayside in Toronto was not quite a greenfield development project, it was one of the most well-known corporate smart cities. In May 2017, Sidewalk Labs, a sister company of Google, won the contract to create a digitally innovative neighbourhood with Waterfront Toronto (WT), a quasi-public agency responsible for the area's real estate development (Flynn and Valverde, 2019). The project aimed to leverage technology and data to establish a low-carbon, pedestrian-oriented community with affordable housing and advanced technologies to improve the quality of life (Flynn and Valverde, 2019; Filion, Moos and Sands, 2023).

However, challenges and controversies plagued the project. For instance, WT lacked expertise in smart city technologies and relied heavily on Sidewalk Labs for this aspect. Concerns about lack of transparency in the planning process were also prevalent, with decisions taken in closed meetings and limited publicly available information (Goodman and Powles, 2019; Sadowski, 2021). Questions were also raised about surveillance technology and the perceived arrogance of Sidewalk Labs in its dealings with the government and the public (Filion, Moos and Sands, 2023). One significant issue was that Sidewalk Labs attempted to use public assets and services and the community as a testing ground to develop profitable technologies without sharing intellectual property (Flynn and Valverde, 2019; Sadowski, 2021).

In May 2020, the discontinuation of the Sidewalk Labs' Quayside project was announced (Spicer and Zwick, 2021). The project's failure also shed light on the challenges and complexities associated with implementing corporate smart city projects, including governance, public engagement, privacy considerations, and striking a balance between private interests and public benefits (Kitayama, 2021; Filion, Moos and Sands, 2023).

12.3.5 Implications from Masdar City, Songdo, Fujisawa, and Quayside

Corporate smart city projects such as Masdar City, Songdo, Fujisawa, and Quayside offer potential solutions for sustainable urban development, but their success hinges on addressing environmental and climate challenges, fostering collaboration, ensuring inclusivity, and prioritising ethical technology use. The development of corporate smart cities has implications pertinent to urban planners and practitioners managing contemporary and forthcoming corporate smart city initiatives. The following implications were selected as focal points for analysis because they emerge as common characteristics among the corporate smart city projects. By examining these implications, the research can offer insights into the fundamental challenges and opportunities that define corporate smart cities.

12.3.5.1 Gradual Progress

Corporate smart city projects may take longer than initially anticipated to fully develop and become operational. Factors like budget constraints, bureaucratic processes, and the need to integrate multiple technologies can slow down progress. It is essential to manage expectations and communicate effectively with the public about the timeline and milestones (Angelidou, 2014; Cugurullo, 2018; Randeree and Ahmed, 2019).

12.3.5.2 Public Scepticism

The stagnation or discontinuation of some corporate smart city projects, such as Quayside in Toronto, can lead to public scepticism about corporate-led initiatives.

Given that concerns about privacy, transparency, and the influence of private interests may undermine public trust, smart city initiatives should prioritise inclusive public engagement, democratic decision-making, and transparent planning processes to gain public acceptance (Cugurullo, 2018; Flynn and Valverde, 2019; Filion, Moos and Sands, 2023).

12.3.5.3 Technology Innovation

Corporate smart city projects serve as living labs and testbeds for various technological innovations. The projects allow companies to develop and commercialise new technologies, ranging from renewable energy solutions and smart infrastructure to surveillance and data analytics. However, ensuring the responsible and ethical use of these technologies is essential to address concerns about privacy and social inclusion (Randeree and Ahmed, 2019; Sgouridis and Kennedy, 2010; Tokoro, 2018).

12.3.5.4 Sustainable Urban Development

The focus on sustainability in smart city projects reflects the growing awareness of environmental concerns. Projects aim to reduce carbon emissions, promote renewable energy sources, and adopt eco-friendly practices. However, striking a balance between environmental goals and economic viability remains crucial to achieving long-term sustainability (Barrett, DeWit and Yarime, 2021; Cugurullo, 2018; Griffiths and Sovacool, 2020).

12.3.5.5 Energy and Climate Insights

The dawn of smart cities has brought to the forefront projects, each grappling with the nuances of integrating technology with sustainability and climate objectives (Griffiths and Sovacool, 2020; Cugurullo, 2018; Randeree and Ahmed, 2019; Kuecker and Hartley, 2019). For instance, Masdar City's recalibrated carbon-neutrality goals and energy challenges (Sgouridis and Kennedy, 2010) and FSST's smart energy systems within Panasonic's vision (FSST Management, 2023) reflect smart city aspirations with the realities of energy and climate imperatives. The experiences of these smart cities reinforce the necessity of a multi-faceted approach to energy and climate issues.

12.4 Results

Our investigation into Woven City reveals a forward-looking enterprise by which Toyota is reshaping urban living with advanced mobility and technology. However, this transition is not without its complexities, as the project navigates between corporate governance and community engagement. As a microcosm of innovation, Woven City embodies a corporate determination to harness technology for societal benefit, yet it must reconcile the scope of public involvement with its top-down implementation approach.

12.4.1 Findings from Document Analysis

The advancement of smart cities requires a convergence of city planning, technological innovation, and corporate willingness to create futuristic, efficient, and sustainable habitats. The findings, drawn from the document analysis, shed light on the origins, objectives, structure, and operational systems of Woven City. Woven City is an ambitious initiative that was first announced by Toyota's former president, Akio Toyoda (2009–2023), at the Consumer Electronics Show (CES) 2020. The project aims to create a smart city where various technologies are integrated, such as automated driving, mobility as a service (MaaS), smart homes, and artificial intelligence (AI) (Sakuma and Kato, 2020).

The Woven City layout features a network of three types of streets with distinct functions: one dedicated to self-driving and eco-friendly vehicles; another thoughtfully configured to accommodate a blend of slower modes of mobility alongside pedestrian traffic, and finally, broad thoroughfares exclusively reserved for pedestrian use. The nomenclature of this city, Woven, is inspired by the fabric of its streets, artfully intertwined to create a harmonious urban landscape (Sakuma and Kato, 2020).

Woven City has been planned and developed with WbyT, the mobility technology subsidiary of Toyota Motor Corporation. Its mission is to deliver safe, intelligent, human-centred mobility for all, through Woven City – a test course for advanced mobility (Toyota, 2023a). WbyT's vision focuses on extending the value of the car, expanding mobility into new realms, and integrating mobility with social systems. With this framework, WbyT is helping Toyota realise a mobility society in which everyone can move freely, happily, and safely (Woven by Toyota, 2023).

WbyT was introduced in April 2023 as a renamed version of Woven Planet Holdings, which comprised the former Woven Core and Woven Alpha (effective from January 2021). Woven Capital remains as an investment arm of WbyT. Woven Core was responsible for the development of automated driving technology and advanced driver assistance systems. Woven Alpha aimed to expand its business through innovative projects such as Woven City. Woven Capital was a global investment fund, which invested in growing companies that are developing innovative technologies. Woven Planet Holdings originated as the Toyota Research Institute-Advanced Development (TRI-AD) established in Tokyo in March 2018. TRI-AD was formed to provide fully integrated, production-quality software for automated driving. Its mission was to create world-class technology and build safe mobility products for the world (Toyota, 2023b). Daisuke Toyoda, the son of Akio Toyoda, serves as the senior vice-president of these organisations.

Akio Toyoda, the former president of Toyota, claimed, "we will work with the local community, step by step, towards the future" and "cooperate in the development of the Iwanami Station [the nearest station to Woven City]" (Toyota Times,

2021). Nonetheless, Woven City is envisioned as a corporate smart city, where Toyota controls the design and implementation of technological infrastructure, which has resulted in limited local citizen participation in the decision-making process (Kitayama, 2021).

On "Toyota Times" (2021), an online platform and publication established by Toyota to serve as a source of news, insights, and stories related to Toyota's innovations and corporate activities, Akio Toyoda claimed:

> Woven City is a test course for Toyota to become a mobility company. This experiment can only be conducted in the framework of building of a "city". However, this city will never be completed, as the goal is not to bring a product to completion. To be honest, there is no need to explain every single detail about Woven City. We have no intention to make the process democratic. I want to take the first step in a closed forum to be dictated by Toyota Motor Corporation, using its own land and own funds without using any tax money of the city of… It should be a top-down decision-making process in the initial phases.

The website of Woven City (2023a) further elaborates that in its system, developers from various companies including Toyota, start-ups, entrepreneurs, and residents, work together to mass-produce future happiness for others – specifically, for someone other than themselves. Sakichi Toyoda, the founder of Toyota, invented the automatic loom with a single thought in mind: to lighten the burden of weaving for his mother. This principle – make someone happy – is the starting point of Toyota, and it remains unchanged even today. It is also the inspiration for the name Woven, which references weaving. Toyota has transitioned from being an automobile manufacturer to a mobility company, guided by the vision "Mobility for All", with an aim to transcend the conventional concept of mobility to bring freedom and enjoyment of mobility to people of all backgrounds. Toyota, along with its various partners, will participate in real-life scenarios and collaborate to shape the future (Woven City, 2023b, 2023c).

Notably, Woven City serves as a premier testing ground for innovative experiments. For example, it facilitates the trial of novel mobility systems that have the potential to lower traffic accident rates and provides a platform for collecting user feedback on newly developed applications (Woven City, 2023c).

12.4.2 Findings from Interview Studies

In the evolving landscape of urban development, Woven City distinguishes itself as an example of innovation and sustainability and by its community-centric ethos (despite the ethos, however, the involvement of the local community in the decision-making process has actually been limited, as suggested earlier). By interweaving insights from key stakeholders, including Toyota Group insiders, a FSST staff, a government official, and a university professor, we present a comprehensive picture of Woven City, shedding light on the nuances that define this intriguing urban experiment.

12.4.2.1 How Did the Project Get the Name Woven?

Upon asking about the origin of the term "Woven" in Woven City, one respondent from Toyota Group said, "Woven can refer to the various types of streets that are folded together in Woven City" (T-5). However, he/she continued:

> Rather, the name Woven originates from the automatic loom (weaving machine) that Sakichi Toyoda invented to help his mother. He/she then went on to establish Toyota Industries Corporation, the first Toyota Group entity. The core philosophy of the company since then is "for someone other than oneself".
>
> (T-5)

Thus, Woven City finds its roots in the innovative history of the Toyota Group: it harks back to the automatic weaving machine invented by Sakichi Toyoda, the founder.

12.4.2.2 How Did the Concept of Woven City Originate?

Subsequently, we delved into the origins of the Woven City concept, seeking to unravel the narrative behind its inception. One respondent from Toyota Group reported that the concept was proposed when the decision was taken to close Toyota's East Fuji Plant in the city of Susono in 2018:

> after the Great East Japan Earthquake in March 2011, there was talk of relocating car production plants to Tohoku (the region where the earthquake struck), and the decision was made to close Toyota's East Fuji Plant in Susono in 2018. The then-president Toyoda had a talk with the East Fuji Plant workers. The atmosphere was bleak. One of the workers asked him, "What will happen to our jobs after the plant closes?" He/she continued, "I can go to Tohoku, but some of my colleagues cannot". After a few moments of silence, Toyoda replied, "We will establish a testing ground here for inventions that will benefit people".
>
> (T-5)

He/she explained how the concept of Woven City emerged as a testing ground for the comprehensive exploration of mobility solutions:

> in 2019, Toyota's transition to a mobility company was announced. Just as the loom became an automobile company, Toyota will now become a mobility company. Woven City is a test course for mobility. The word "Move" in "Mobility" refers not only to the movement of people, goods and information, but also to emotions.
>
> (T-5)

While the narrative presented so far serves as the "official explanation" for the genesis and realisation of the concept of Woven City, a Susono city government official offered additional insights:

> about 10 years ago, after the Great East Japan Earthquake, there was talk of Toyota moving its plant to Tohoku, as part of its contribution to the region, and establishing a smart city that would utilise hydrogen and solar power and efficiently run its energy infrastructure, with the plant at its core. I think the idea was to make individual lives smarter, but that story was cut short and the talk turned to creating a "Woven City" in the city of Susono.
>
> (G-1)

The concept of Woven City originated in response to the closure of Toyota's East Fuji Plant in Susono in 2018, which followed discussions about relocating car production plants after the Great East Japan Earthquake in 2011. However, an alternative perspective suggests that discussions about Toyota relocating its plant and establishing a smart city with sustainable energy practices to enhance individual lives with smart technology has been ongoing for a decade. Over time, this vision may have transformed into the concept of Woven City in Susono, demonstrating the complex evolution of city planning and development.

12.4.2.3 Can Woven City Be Considered a Corporate Smart City?

Our subsequent inquiry delved into the perception of Woven City as a corporate smart city. We intended to discern whether those engaged in the project viewed it primarily as a corporate-driven smart city or if there were alternative perspectives that portrayed it differently. One respondent from the Toyota Group noted that Woven City is destined to serve as an experimental smart city: "Woven City will be an experimental smart city with a large early adopter population" (T-2). Another respondent from Toyota Group referred to Woven City as a visionary corporate smart city:

> Woven is a smart city to be built by Toyota, and is designed as a next-generation urban form. This is an advanced experimental approach to smart energy, mobility and lifestyle. Woven is a place with no restrictions on personal information. It will be a long time before the data obtained from within is made available to the outside world. Most of the residents will be researchers from abroad or Toyota employees.
>
> (T-1)

However, one respondent from the Toyota Group provided a perspective that Woven City is a test course: "inside WbyT, Woven City is perceived as a test course rather than a city. It is not that we want to create a city, but we want to know what residents will do there" (T-3). Another respondent from the Toyota Group emphasised that the term "smart city" should not be used within WbyT because Woven City, in their view, is not a typical smart city:

> we are not supposed to use the term "smart city" inside WbyT because Woven City is a test course to accelerate innovation through experiments. A smart

> city is associated with the image of a convenient city with the use of ICT, but Woven City is not one (a convenient city). Toyota is huge, so there may be some misunderstanding about it even among Toyota employees.
>
> (T-4)

Another respondent from WbyT also claimed that Woven City is not classified as a conventional smart city; instead, it is best considered a test course designed to enhance quality of life:

> Woven City is definitely not a smart city: it is a test course for people to live better. It does contain elements of a smart city, and it is a city in the sense that people actually live in it. For us, however, it is a test course in the form of a city, and city development is not our primary objective. Essentially, we want inventors or people with an inventor mindset to live there. We aim to attract people who are eager to contribute to the betterment of others. Our ultimate goal is the mass production of happiness.
>
> (T-5)

The respondent from academia expressed the belief that Woven City is a smart city but highlights a difference in its approach compared with typical smart cities:

> I consider Woven City to be a smart city. However, smart cities typically prioritise addressing specific issues, such as the healthcare needs of the elderly, which can have a global appeal. In contrast, Woven City does not appear to focus on solving specific problems.
>
> (A-1)

We also engaged in dialogue with a member of the FSST team regarding some Toyota personnel's decision not to label Woven City as a smart city. The staff member remarked, "We cannot surmise the reasons behind their reluctance to designate Woven City as a 'smart city'", indicating that the term "smart city" is generally considered to have a positive connotation, similar to a brand enhancement.

These findings suggest that Woven City is multifaceted and may encompass elements of a smart city, but its primary purpose is experimental innovation and improving the well-being of its residents, potentially differentiating it from traditional smart city concepts. The lack of a unanimous definition underscores the evolving nature of this unique project.

12.5 Discussion: Can Woven City Be Considered a Corporate Smart City?

Based on the analyses of the findings from the literature review, document analysis, and interview data, we suggest considering Woven City a type of corporate smart city, as it shares many characteristics with other corporate smart cities identified in the literature review, such as corporate leadership and funding, technological integration, and innovation hubs.

Corporate smart cities such as Woven City, Masdar City, Songdo, Fujisawa, and Quayside are spearheaded, funded, and managed by large corporations. Corporate smart cities often represent the integration of cutting-edge technologies, including AI, automation, smart homes, and advanced mobility such as MaaS (Griffiths and Sovacool, 2020; Tokoro, 2018) – traits also shared by Woven City. All these cities serve as testing grounds or living labs for corporations to experiment with new technologies, solutions, and ideas in a real-world urban setting (Kummitha, 2019; Randeree and Ahmed, 2019; Strickland, 2011; Toyota, 2023a).

However, the interviews revealed that some within Toyota classify Woven City as a conventional smart city (T-2, T-3), while others do not (T-1, T-4, T-5). It may differ from other corporate smart cities in various respects, notably Toyota's complete control and ownership, emphasis on experimental innovation, focus on mobility, unique identity, and potential inconvenience (which will be discussed shortly in this section). Unlike some other corporate smart cities that collaborate with external partners and investors, Woven City is fully controlled by Toyota: the design and implementation of technological infrastructure are under Toyota's authority, limiting local stakeholder participation in decision-making (Kitayama, 2021).

The Woven City project is being conducted on Toyota's privately owned land, funded by the corporation itself, without reliance on tax money. In contrast, the ownership structure, funding mechanisms, and involvement of public funds vary significantly among other corporate smart cities. Toyota's robust ownership is also relevant to the control of privacy or protection of personal data within Woven City. This is because the project is owned by Toyota and only those who provide informed consent to be part of the experiment can enter the city, as some interviewees reported (T-1, T-2, T-3, T-4).

It is also noteworthy that although most corporate smart cities prioritise technology and efficiency, Woven City has a particular emphasis on mobility with a particular focus on automated driving and AI, as it is backed by Toyota. Although other smart cities such as Masdar City and FSST also incorporate mobility as a significant element, Toyota's focus on mobility distinguishes Woven City from other corporate smart cities that may prioritise different aspects of urban living.

Another notable difference is that Toyota seems to deliberately distance Woven City from the smart city label, emphasising that it is a test course in the form of a city. This is what lends it a unique identity, unlike other corporate smart cities that embrace the smart city label. As one respondent from the Toyota Group (T-4) said, "Woven is not a smart city, which is typically associated with convenience".

Woven City diverges from the standard smart city model by primarily serving as a test course for experiments to facilitate the mass-production of happiness; the emphasis on experimental approach may lead to occasional inconvenience for residents or visitors as part of the pursuit of this overarching objective. To underscore our perspective: Woven City ultimately aims to be a convenient city – whether consciously or subconsciously – as a means to improve quality of life because convenience is a fundamental aspect of the city's mission – the mass production of happiness – albeit not its sole focus.

In the ambit of sustainability, Woven City exemplifies a deliberate effort to synergise technological progress with environmental stewardship. Toyota's investment in automated driving and AI is complemented by the adoption of renewable energy systems and eco-friendly materials, positioning the project in line with broader global objectives of energy efficiency and climate action. The city's design anticipates a low-carbon future, leveraging Toyota's expertise in hydrogen fuel cells and renewable energy sources to reinforce its role as a paradigm of eco-conscious urbanism. By prioritising environmental considerations alongside technological innovation, Woven City sets out to redefine corporate smart cities, aspiring not only innovation hubs but vanguards of sustainable urban life.

12.6 Conclusion

This chapter contributes to the discourse on contemporary smart city development showcasing Toyota's Woven City project as a case study that integrates corporate innovation with a steadfast commitment to sustainability, energy efficiency, and climate responsiveness. Diverging from traditional smart city models, Toyota's Woven City exemplifies a novel approach by merging its mobility prowess with environmental foresight. This alignment results in a unique urban fabric where top-down decision-making and technological innovation contribute to the shaping of the city's development in harmony with sustainability.

While Woven City may diverge from the conventional smart city label, opting instead for an experimental ethos, its essence lies in fostering an improved quality of life within a sustainable framework. It aspires to establish new standards for urban living that prioritise technological innovation and a concerted effort to reduce carbon emissions, promote clean mobility solutions, and support a transition towards a low-carbon economy. Ultimately, Woven City's narrative is a transformative urbanism that resonates with the global imperative of crafting cities that are both innovative and attuned to the exigencies of our climate, positioning Toyota at the vanguard of shaping smart, sustainable futures.

Woven City presents a bold vision of urban development where technology meets sustainability. This conclusion reaffirms the project's alignment with environmental stewardship through the integration of clean mobility solutions and sustainable energy practices. Emphasising energy efficiency and reducing carbon emissions are pivotal in the city's design, reflecting a commitment to combat climate change. Toyota's initiative not only pushes the envelope in automotive technology and urban living but also stands as a proactive response to global sustainability challenges, potentially setting a precedent for future smart city endeavours to follow a similar environmentally conscious path.

Note

1 In Arabic: مصدر.

References

Angelidou, M. (2014) "Smart City Policies: A Spatial Approach", *Cities*, 41, pp. 3–11.

Barrett, B.F.D., DeWit, A. and Yarime, M. (2021) "Japanese Smart Cities and Communities: Integrating Technological and Institutional Innovation for Society 5.0", in H.M. Kim, S. Sabri A. Kent (eds), *Smart Cities for Technological and Social Innovation: Case Studies, Current Trends, and Future Steps*. Cambridge, MA: Academic Press, pp. 73–94.

Benedikt, O. (2016) "The Valuable Citizens of Smart Cities: The Case of Songdo City", *Graduate Journal of Social Science*, 12(2), pp. 17–36.

Bowen, G.A. (2009) "Document Analysis as a Qualitative Research Method", *Qualitative Research Journal*, 9(2), pp. 27–40.

Caputo, F., Buhnova, B. and Walletzky, L. (2018) "Investigating the Role of Smartness for Sustainability: Insights from the Smart Grid Domain", *Sustainability Science*, 13(5), pp. 1299–1309.

Cardullo, P. and Kitchin, R. (2019) "Smart Urbanism and Smart Citizenship: The Neoliberal Logic of 'citizen-focused' Smart Cities in Europe", *EPC: Politics and Space*, 37(5), pp. 813–830.

Cugurullo, F. (2018) "Exposing Smart Cities and Eco-Cities: Frankenstein Urbanism and the Sustainability Challenges of the Experimental City", *Environment and Planning A: Economy and Space*, 50(1), pp. 73–92.

Dameri, R.P. (2017) *Smart City Implementation*. Cham: Springer.

Filion, P., Moos, M. and Sands, G. (2023) "Urban Neoliberalism, Smart City, and Big Tech: The Aborted Sidewalk Labs Toronto Experiment", *Journal of Urban Affairs*, 45(9), pp. 1625–1643.

Flynn, A. and Valverde, M. (2019) "Planning on the Waterfront: Setting the Agenda for Toronto's 'Smart City' Project", *Planning Theory & Practice*, 20(5), pp. 769–775.

FSST Management (2023) *FujisawaSST*. Fujisawa: Fujisawa SST Council.

Goodman, E. P. and Powles, J. (2019) "Urbanism under Google: Lessons from Sidewalk Toronto", *Fordham Law Review*, 88(2), pp. 457–498.

Griffiths, S., and Sovacool, B.K. (2020) "Rethinking the Future Low-Carbon City: Carbon Neutrality, Green Design, and Sustainability Tensions in the Making of Masdar City", *Energy Research & Social Science*, 62, pp. 1–9.

Holguín Rengifo, Y.X., Herrera Vargas, J.F. and Valencia-Arias, A. (2023) "Proposal for a Comprehensive Tool to Measure Smart Cities under the Triple-Helix Model: Capacities Learning, Research, and Development", *Sustainability*, 15, pp. 1–21.

Hornyak, T. (2022) "Why Japan Is Building Smart Cities from Scratch", *Nature*, 608, pp. 32–33.

Janajreh, I., Su, L. and Alan, F. (2013) "Wind Energy Assessment: Masdar City Case Study", *Renewable Energy*, 52, pp. 8–15.

Japan Electrical Manufacturers' Association. (2016). *Shipment Performance of Industrial General-Purpose Electrical Equipment*. In Japanese: 産業用汎用電気機器の出荷実績 [Sangyōyō Hanyō Denki Kiki no Shukka Jisseki]. Available at: www.jema-net.or.jp/Japanese/data/2016/16.07/1607jds-comment.pdf

Kitayama, S. (2021) *Automotive-driven City of Smartness: Study of Compatibility Between Toyota's Woven City and its Current Business Model*. Master Thesis, Keio Business School: Tokyo.

Kshetri, N. and Alcantara, L. and Park, Y. (2014) "Development of a Smart City and Its Adoption and Acceptance: The Case of New Songdo", *Digiworld Economic Journal*, 96(4), pp. 113–128.

Kuecker, G.D. and Hartley, K. (2019) "How Smart Cities Became the Urban Norm: Power and Knowledge in New Songdo City", *Annals of the American Association of Geographers*, 110, pp. 516–524.

Kummitha, R.R.K. (2019) "Smart Cities and Entrepreneurship: An Agenda for Future Research", *Technological Forecasting & Social Change*, 149, pp. 1–10.

Matsumoto, T., Esaka, T. and Izumiyama, R. (2023). "A study on Initiatives Related to Regional Characteristics of Smart Cities". In Japanese: スマートシティの地域特性における取り組みに関する研究 [Sumāto Shiti no Chiiki Tokusei ni Okeru Torikumi ni Kansuru Kenkyū]. Reports of the City Planning Institute of Japan, 21, 351-356.

Morgan, H. (2022) "Conducting a Qualitative Document Analysis", *The Qualitative Report*, 27(1), pp. 64–77.

Mullins, P.D. and Shwayri, S.T. (2016) "Green Cities and 'IT839': A New Paradigm for Economic Growth in South Korea", *Journal of Urban Technology*, 23(2), pp. 1–18.

Panasonic. (2007). *The Panasonic Report for Sustainability 2007*. Available at: https://holdings.panasonic/global/corporate/sustainability/pdf/csr2007.pdf

Panasonic. (2016). *Annual Report 2016*. Available at: https://www.annualreports.co.uk/HostedData/AnnualReportArchive/p/NYSE_PC_2016.pdf

Randeree, K. and Ahmed, N. (2019) "The Social Imperative in Sustainable Urban Development: The Case of Masdar City in the United Arab Emirates", *Smart and Sustainable Built Environment*, 8(2), pp. 138–149.

Sadowski, J. (2021) "Who Owns the future city? Phases of Technological Urbanism and Shifts in Sovereignty", *Urban Studies*, 58(8), pp. 1732–1744.

Sakuma, D. and Kato, K. (2020) "The Background of the Automotive Industry's Smart City Strategy and the Pros and Cons of Winner Takes All (WTA): Insights from the Case of Toyota's Woven City." In Japanese: 自動車業界のスマートシティ戦略の背景とWTA（Winner takes all: 勝者総取り）の功罪 [Jidōsha gyōkai no sumāto shiti senryaku no haikei to WTA (Winner takes all: shōsha sōdori) no kōzai]. *Development Engineering*, 40(1), pp. 37–40.

Sakurai, M. and Kokuryo, J. (2018) "Fujisawa Sustainable Smart Town: Panasonic's Challenge in Building a Sustainable Society", *Communications of the Association for Information Systems*, 42, pp. 508–525.

Scuotto, V., Ferraris, A. and Bresciani, S. (2015) "Internet of Things; Applications and Challenges in Smart Cities: A Case Study of IBM Smart City Projects", *Business Process Management Journal*, 22(2), pp. 357–367.

Segel, A. *et al.* (2012) "New Songdo City", *Harvard Business School Case*, 9, pp. 206–219.

Selinger, M. and Kim, T. (2015) "Smart City Needs Smart People: Songdo and Smart + Connected Learning", in D. Araya (ed.), *Smart Cities as Democratic Ecologies*. New York, NY: Palgrave Macmillan, pp. 159–172.

Sgouridis, S. and Kennedy, S. (2010) "Tangible and Fungible Energy: Hybrid Energy Market and Currency System for Total Energy Management: A Masdar City Case Study", *Energy Policy*, 38(4), pp. 1749–1758.

Singh, A. and Singla, A.R. (2020) "Constructing definition of Smart Cities from Systems Thinking View", *Kybernetes*, 50(6), pp. 1919–1960.

Spicer, Z. and Zwick, A. (2021) "A Smart City for Toronto: What Does Quayside Tell us about the State of Smart City Building", in Z. Spicer and A. Zwick (eds) *The Platform Economy and the Smart City: The Transformation of Urban Policy and Governance*. Montreal: McGill-Queen's University Press, pp. 249–265.

Strickland, E. (2011) "Cisco Bets on South Korean Smart City", *IEEE Spectrum*, 48(8), pp. 11–12.

Tokoro, N. (2018) "Realization of a Health Support Ecosystem Through a Smart City Concept: A Collaborative Dynamic Capabilities Perspective", in M. Kodama (ed.), *Collaborative Dynamic Capabilities for Service Innovation: Creating a New Healthcare Ecosystem*. Cham, Switzerland: Palgrave Macmillan, pp. 135–151.

Toyota (2023a) "Woven by Toyota to Accelerate Toyota's Vision for Mobility". Available at: https://global.toyota/en/newsroom/corporate/39070846.html (Accessed: 30 March 2024).

Toyota (2023b) "Toyota Research Institute – Advanced Development Inc. (TRI-AD) Announces It Will Expand and Its Operations by Forming Woven Planet Holdings and Two New Operating Companies, Woven Core and Woven Alpha". Available at: https://global.toyota/en/newsroom/corporate/33786527.html (Accessed: 30 March 2024).

Toyota Times (2021) "Woven City Started: What President Toyoda Promised at the Groundbreaking Ceremony". In Japanese: Woven Cityはじまる 豊田社長が着工式で約束したこと [Woven City hajimaru Toyoda shachō ga chokkōshiki de yakusoku shita koto]. Available at: https://toyotatimes.jp/toyota_news/construction_wovencity/120.html (Accessed: 30 March 2024).

Woven by Toyota (2023) "What We Build". Available at: https://woven.toyota/en/what-we-build/ (Accessed: 30 March 2024).

Woven City (2023a) "Above-Ground and Underground Mobility". Available at: www.woven-city.global/ (Accessed: 30 March 2024).

Woven City (2023b) "Woven City Press. Vol. 1". Available at: www.woven-city.global/ (Accessed: 30 March 2024).

Woven City (2023c) "Woven City Press. Vol. 2". Available at: www.woven-city.global/ (Accessed: 30 March 2024).

Yin, R. (2018) *Case Study Research and Applications: Design and Methods.* Thousand Oaks Trees, CA: SAGE Publications.

13 Earth Observation for Digital Twin

Fusion of Smart City and Satellite Data for Japan's Energy Transition

Naoko Sugita, Naoko Matsuo, Takuji Kubota, and Misako Kachi

13.1 Introduction

This paper focuses on the recent trends and efforts towards realising smart cities, with possible use and integration of satellite Earth observation (EO) data in such efforts. A smart city, in Japanese context is defined as:

> a sustainable city or region that continues to produce new values by solving issues through updating management (planning, maintenance, management and operation) utilising new technologies such as Information and Communication Technology (ICT), and thus the leading test bed for Society 5.0.
>
> (Cabinet Office, Government of Japan, 2024)

Earth observation (EO) satellite data is a vital resource to understand global environmental change and is crucial for scientific research addressing global scale issues. The benefit deriving from EO data in the monitoring environment should be helpful in effective planning and operation of smart cities. Japan is among the limited number of countries that possess space technologies that generate EO data, and also is a country scarce of energy resources. These are the reasons why she should make full use of EO data and information obtained from such data.

The Group on Earth Observations (GEO) summarised such benefits of satellite EO with polar orbiting, sun synchronous EO sensors that allow observation in "wide swaths of the Earth in one pass". These are regular and repeatable observations that allow:

> consistent and systematic surface observations of the entire Earth surface, multi-annual time series of observations (some dating back to the 1970s), and cost-effective means for monitoring remote and inaccessible areas; with some exceptions around polar regions, EO satellites can observe any location on the Earth's surface at some time in their orbit, which many are not possible by ground-based surveys.

EO data and data obtained by airborne, ocean-based, and *in situ* sensors complement each other; while satellite EO allows stable acquisition of global data, *in situ*

DOI: 10.4324/9781003471417-13

sensors provide data for remote sensing and modelling with important verification. Satellite EO could also overcome difficulties and restrictions associated with national boundaries, which are specifically helpful for environmental indicators (Anderson *et al.*, 2017).

Environmental monitoring that shall serve as the basis of smart city aiming for efficient use of energy would benefit from a wide range of satellite data such as temperature, ice extent, sea level, land cover change. Currently, specific skills are usually necessary to access or use this data (Sugita *et al.*, 2023, p. 1), however they are useful to improve lives in cities and solve societal issues. It is in this regard to explore possible contributions of EO satellite data in smart cities. In this context, this chapter discusses recent examples of satellite EO data being applied to enhance understanding of various phenomena, and how this EO data can be integrated into larger data platforms, as well as the state and recent achievements in visualising various physical quantities achieved by satellite EO data.

13.2 Smart City and Energy Transition

Without doubt, smart solutions regarding energy are the key for any sustainable smart city. According to the *Smart City Guidebook*, the realisation of a city or region with low environmental impact is one of the main objectives. It is expected that energy and resources are optimally utilised in any scenes of city lives, aligned with the real movements of people and goods, thus effective in realising a decarbonised society (Cabinet Office, Government of Japan, 2023, p. 11).

To a certain extent, the digital twin of the city would make simulation of energy efficiency possible. Furthermore, inclusion of EO satellite data in such a digital twin would make the simulation more precise, thereby making the city more efficient. Renewable energy, such as solar power utilising cloud data and offshore wind power generation, would become more efficient. Frequent monitoring of activities in cities and infrastructure, as well as realising the digital twin of the city should make activities in the city automated and unmanned, resilient and environmentally less impactful, thus becoming essential in realising a smart city.

13.3 City Digital Platform and Data Fusion

One example of possible contribution of EO satellite data in Japan for smart cities to be explored is the city digital platform and data fusion. When utilising EO data for smart city, the basis would be digital platform and EO data combined with other data sources. In this regard, PLATEAU project, a city digital twin led by the Ministry of Land, Infrastructure, Transport and Tourism (MLIT), and GOBLEU project, an early attempt to combine satellite and airborne data measuring greenhouse gases (GHGs), would be studied for possible future application.

These are two examples, digital platform basis and environment monitoring information, which would be essential for the operation of a smart city. City digital platforms that will increasingly use satellite data, and other examples in Fukuoka combining satellite data and other data sources for more accurate and comprehensive understanding of GHG distribution, are two examples of recent trends and efforts in using satellite EO data for addressing societal issues. While such applications are still at nascent stage, efforts to integrate satellite EO data should be pursued to make lives better in both cities and suburban areas.

13.3.1 PLATEAU

MLIT is leading the project PLATEAU, started in 2020, to develop and open data on 3D city models throughout Japan as city digital twin. Participated in by local governments, private companies, researchers, engineers, and creators, the project aims to promote use of 3D city models and to realise digital transformation (DX) of urban policy, and the platform has been growing since its inception. It aims to address societal issues by digital transformation of the cities (MLIT, 2024). When this 3D city model is used in public, the Japan Aerospace Exploration Agency (JAXA) expects satellite EO data would be utilised for periodic updates. Furthermore, by integrating simulation based on EO data, it is expected that many use cases of smart city would occur.

13.3.2 GOBLEU Project

Greenhouse Gas Observations of Biospheric and Local Emissions from the Upper sky (GOBLEU) is a project undertaken by JAXA, Japanese space agency, and All Nippon Airways (ANA), one of the major national Japanese carriers (EORC and JAXA, 2024a). Dedicated to observing the concentration distribution of GHGs from outer space, the Greenhouse Gases Observing Satellite (GOSAT) has been observing GHGs since its launch in 2009. In this research project with ANA, JAXA has applied the remote sensing technology with the spectrometer onboard GOSAT to passenger aircrafts operated by ANA. Since October 2020, JAXA and ANA have collaborated in measuring horizontal distribution of atmospheric compositions. Carbon dioxide (CO_2) and nitrogen dioxide (NO_2) over megacities and plant sun-induced chlorophyll fluorescence (SIF) are measured with much higher spatial resolution from the upper sky. Analytic results are made public on the web (EORC and JAXA, 2024b).

By combining the satellite data and results of this GOBLEU study, estimation has been made on local emissions from different source sectors such as power plants, industries, and transport. Data was first acquired by regular flights between Tokyo and Fukuoka city, with the expectation to obtain emission data over urban areas that are responsible for more than 70% of future global total GHG emissions. In this regard, it would contribute to the Paris Agreement by presenting reduction effects from different source sectors (EORC and JAXA, 2024a).

13.4 Possible Future Use Cases of Satellite EO Data for Smart Cities

In the previous section, we discussed some early applications of satellite EO data integrated into the data platform and examples of EO data combined with other sources. The evolution of such applications and possible future use cases of satellite EO data for smart cities are anticipated. For example, the Consortium for Satellite Earth Observation (CONSEO), a Japanese industry–academia–government community established in 2022 to promote EO data application by providing inputs to policy process, has identified following four areas as possible contribution related to smart city efforts (CONSEO, 2024, p. 18).

According to CONSEO, the first is monitoring, responding, and optimising activities in city and city infrastructures. It offers a contribution to digital transformation (DX) for disaster prevention, agriculture and forestry, construction and civil engineering, and cartography. Second, optimisation of urban planning that enables monitoring land use and assessing policies by indicators. Third, creating and updating city digital twin, which is AI learning space for automated driving and policy studies by simulations. Finally, monitoring urban environment and visualising environmental value enables estimating suction and emission amount of GHGs in urban areas and visualising natural capital.

13.5 Addressing Climate Change by EO Data

Climate change is one of the key issues that should be considered when pursuing and realising a smart city. Recent abnormal weather and associated disasters force us to take these factors into account for any smart city planning. Following are some examples that EO data may provide solutions to and be useful.

13.5.1 Climate Change and Natural Disasters

As a result of climate change, it is now a shared view that natural disasters such as heavy rainfall are increasing (IPCC, n.d.). Post-disaster measures by the public sector are becoming increasingly important. It is possible to observe night-time and cloudy situations with a wide range by satellites, which comes out as strength of satellite observation that allows early detection of disaster. Local governments could use satellite EO data for initial decision-making. It is this wide coverage of observation by satellites that should be combined and referred to by *in situ* data (Anderson *et al.*, 2017), which is one of the directions smart cities should pursue in its administration.

In Japan, at a national level, Shared Information Platform for Disaster Management (SIP4D) provides processed and analysed data to maximise and utilise existing information to prevent disasters (National Research Institute for Earth Science and Disaster Resilience, 2024).

13.5.2 Today's Earth

Furthermore, as a specific example, JAXA developed the global terrestrial hydrological simulation system called Today's Earth (TE) that integrates satellite observation data, land surface model, and river routing model under collaboration with the University of Tokyo (Ma *et al.*, 2021; Yoshimura *et al.*, 2008). It aims to

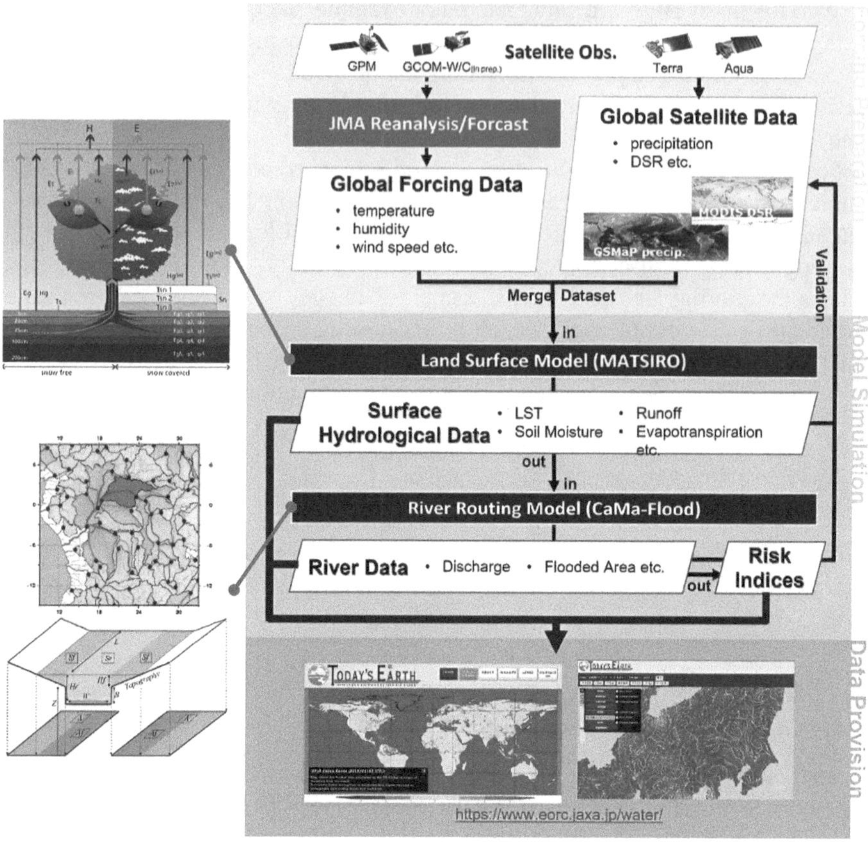

Figure 13.1 System structure of TE.

Source: EORC/JAXA (2024c).

contribute to disaster monitoring, prediction, and water resource management, by simulating over 50 hydrological parameters and risk indicators related to water and land, making the maximum use of satellite observations (Figure 13.1).

Current TE system consists of:

land surface model MATSIRO (Minimal Advanced Treatments of Surface Interaction and Runoff) and river routing model CaMa–Flood (Catchment-based Macro-scale Floodplain). MATSIRO simulates the water and energy interactions between a land surface with a vegetation canopy and atmosphere. Based on the calculated runoff amount, CaMa–Flood enables hydrodynamic simulation with floodplain on a global scale. The model solves the local inertial equation, considering a rectangular river channel and trapezium flood plain storage, and represents flood plain dynamics assuming that the elevation profile of the flood plain monotonically increases in each pixel.

(EORC and JAXA, 2024c)

The TE system targets seamless downscaling from the global scale to the municipality scale to predict the flooding from large rivers to municipal-sized rivers in the world. For example, the TE–Japan system, regional version around Japan with 1 km spatial resolution, uses meteorological forecasts by Japan Meteorological Agency (JMA) Meso-Scale Model (MSM) as forcing to predict future status of hydrological parameters, including river discharge and flood plain, to estimate flood risks more than 30 hours in advance. The TE–Japan predictions are currently provided to the local governments and commercial companies under collaboration activities to examine the possibility of operational utilisation, especially for disaster reaction before the event.

Future projection by simulation with various hydrological parameters would make early evacuation in cases such as heavy rain and linear precipitation belt, which would contribute to building a resilient smart city. By identifying high-risk areas, it would contribute to city planning with efficient use of energy.

13.5.3 Air Quality

The World Health Organization (WHO) defines air pollution as "contamination of the indoor or outdoor environment by any chemical, physical or biological agent that modifies the natural characteristics of the atmosphere" (WHO, 2024). WHO points out that 12% of the global burden of disease is caused by environmental risks, with air pollution ranking first. WHO thus concludes that in order to understand a population's exposure to risks and to take necessary actions, monitoring air quality would be the first step (WHO, 2023).

Common sources of air pollution are motor vehicles, industrial facilities, mineral dust, and forest fires (WHO Africa, n.d.). Aerosols floating in the air, such as yellow dust and PM2.5, can be observed from space. In fact, by incorporating the observation result of aerosols into the simulation, prediction of yellow dust has been improved. PM2.5 particles are known to pose a significant health risk, by travelling deeply into the lungs. Many cities in the world face high concentrations of PM2.5, increasing the risk such as lung cancer and chronic and acute respiratory diseases (World Economic Forum, 2017).

Global Change Observation Mission–Climate (GCOM-C), SHIKISAI, is a satellite launched by JAXA in 2017 to observe physical quantities related to global aerosols. GCOM-C has multiband from near-UV to infrared and relatively high frequency, which allows global observation once every two or three days. By assimilating GCOM-C data into the model, it is expected to improve prediction accuracy.

Figure 13.2 is one example of validation results, showing the horizontal distribution of soil particles and the vertical distribution of aerosols, showing changes and improvements by using assimilation. Aeolian dust information (JMA, 2024) of

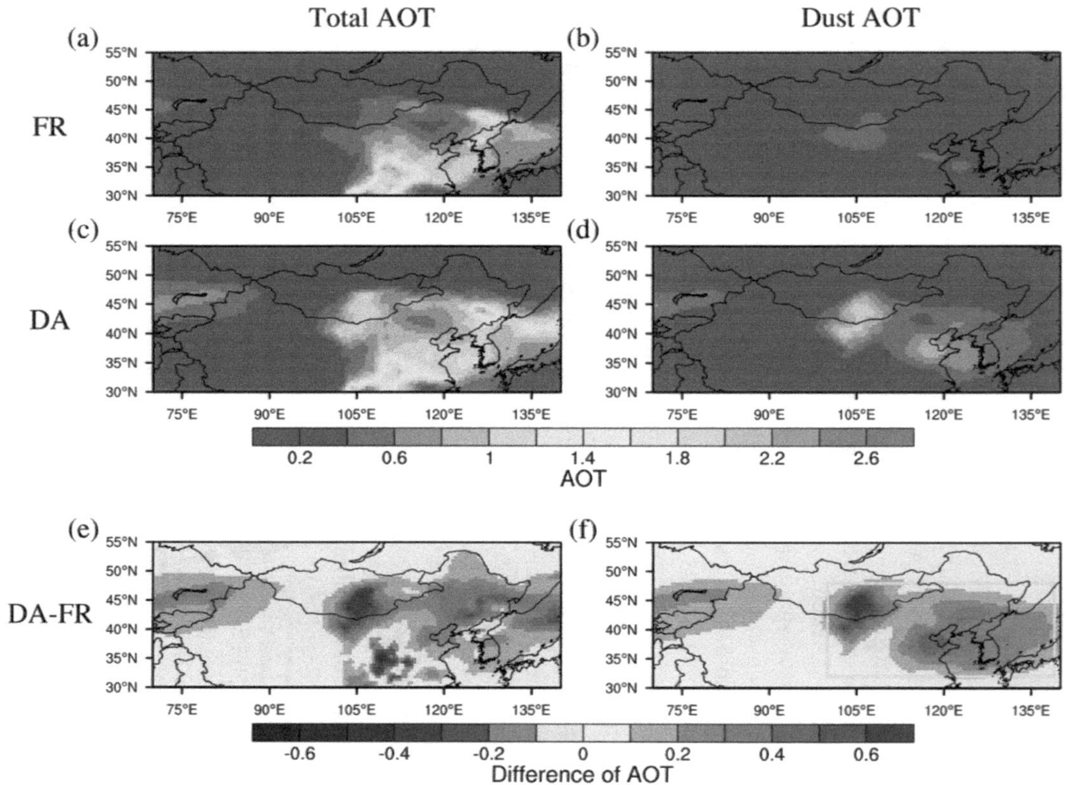

Figure 13.2 Assimilation experiments of NICAM-Chem and Second generation Global Imager (SGLI) aerosol data. (a)-(f): Difference in the horizontal distributions of aerosol optical thickness with/without SGLI assimilation in NICAM-Chem; *Figures on next page*: Comparison of the vertical distributions of aerosol extinction coefficient using ground-based lidar observation.

Source: Cheng *et al.* (2021).

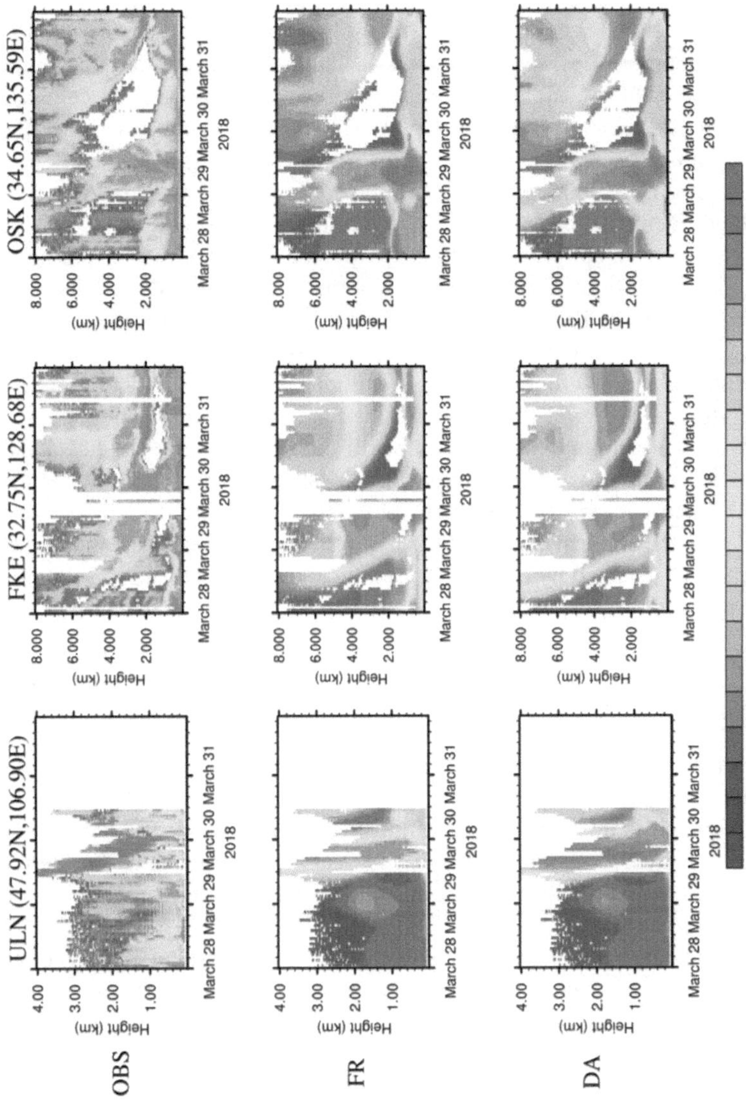

Figure 13.2 (Continued)

the JMA uses some of the findings and algorithms obtained from these assimilation experiments of satellite observation data and models (Yoshida *et al.*, 2018, 2021; Yumimoto *et al.*, 2017, 2018).

13.5.4 Land Surface Temperature

Recent trend of global warming would likely have severe impacts on the life of people living in large cities, especially by heat island effects. Land surface temperature (LST) observed by satellites provides useful information when considering heat island effects. For example, GCOM-C allows monitoring of LST at 250 m resolution, with its Second-generation Global Imager (SGLI). It conducts surface and atmospheric measurements related to the carbon cycle and radiation budget, such as clouds, aerosols, ocean colour, vegetation, and snow, and ice.

Following is an example of detecting local phenomena that is possible by obtaining LST with high (i.e. 250 m) resolution. Figure 13.3 shows the thermal belt on the slope of Mt. Tsukuba in Ibaraki prefecture, Japan. In the daytime (upper left in Figure 13.3), the top of the mountain shows the lowest land surface temperature, and higher at its foot. In contrast, at night (the upper right in Figure 13.3), the temperature is higher at the mountainside than at the top and even compared with the foot of the mountain. This example of high-temperature area is called the "thermal belt". In this case, this is caused by the specific physiographic feature of Mt. Tsukuba, being an independent peak facing the Kanto Plain (Figure 13.4).

Considering the application for smart city, following is an example in the greater Tokyo area, LST detected by GCOM-C (*a*), of the heat in 2018 and its relationship with the vegetation (*b*). This example shows that areas with forests and parks are cooler.

13.5.5 Wildfires

The recent increase of wildfires around the globe caused by climate change poses significant risk to people including those living in cities. According to WHO, wildfire smoke, which is a mixture of hazardous air pollutants, such as PM2.5, NO_2, ozone, aromatic hydrocarbons, or lead, have direct impacts on health. Furthermore, wildfires also impact the climate by releasing into the atmosphere large quantities of carbon dioxide and other GHGs (WHO, n.d.).

WHO points out that wildfire events are getting more extreme in terms of acres burned, duration and intensity, which lead to the disruption of transportation, communications, water supply, and power and gas services (WHO, n.d.). While Japanese cities have not been directly affected by wildfires or related smoke, it is pointed out that climate change has enhanced the drying of organic matter in forests, and thus the number of large fires has doubled between 1984 and 2015 in the western United States (Centre for Climate and Energy Solutions, 2024).

Here, the example of California wildfires that occurred on 16 August 2020, considered to be one of the worst cases in terms of the scale of damage, shows the effort by JAXA to visualise wildfires by satellite observation (JAXA, 2020). With

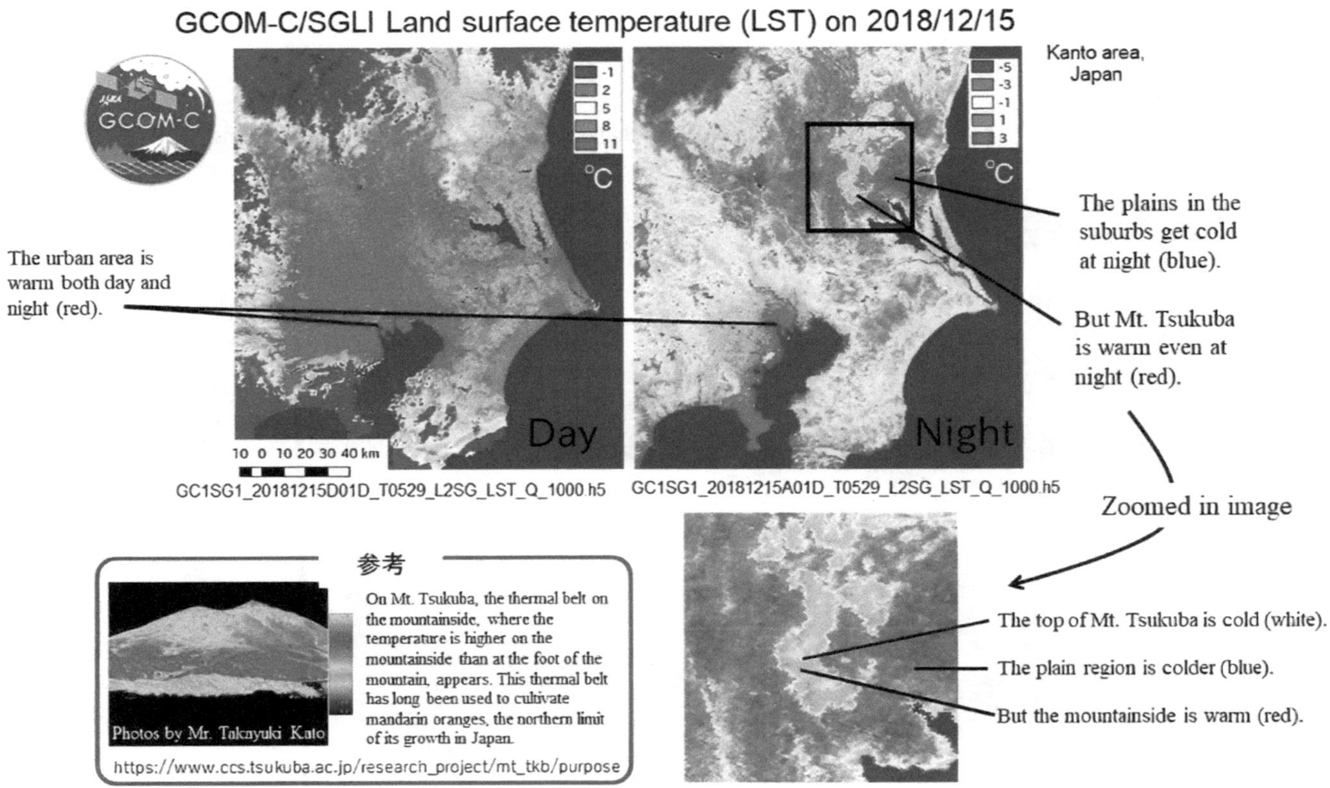

Figure 13.3 Thermal belt detected by GCOM-C on the slope of Mt. Tsukuba.

Source: Nasahara (2022).

Figure 13.4 (a) Land surface temperature detected; (b) vegetation observed by GCOM-C.

Source: JAXA (2018).

the help of satellite data, exceptionally dry conditions of the land and heat waves show that wildfires could easily occur (JAXA, 2020).

13.5.5.1 Drought

California, with its Mediterranean climate, usually has a rainy season from November to winter around March. However, in 2020, California faced record drought with few rains during January to February (NIDIS, 2021).

Utilising multiple satellite data such as the Global Precipitation Measurement (GPM) and Global Change Observation Mission–Water (GCOM-W) SHIZUKU, JAXA generates and provides global precipitation data called Global Satellite Mapping of Precipitation (GSMaP). GSMaP website found at JAXA's Earth Observation Research Centre (EORC) provides the statistical index related to drought, the Standardised Precipitation Index (SPI), calculated by GSMaP. Compared with ground observation, satellites can monitor the condition of a broader area. It is this strength of satellite observation that may contribute to monitoring and may be able to detect anomalies before or at an early stage of negative events such as wildfire that may significantly affect lives of many in cities.

Figure 13.5 is the result from GSMaP of SPI monthly distribution in California region in February 2020. Figure 13.5 shows that SPI was less than –1.5 in most parts of California. SPI value from –1.5 to –2.0 indicates that the condition is severely dry. SPI value of less than –2.0 indicates that it is extremely dry. This result shows that California in 2020 faced significant drought conditions compared with the normal year. It is assumed that forests, vegetation, and its soil in California had been in exceptionally dry condition.

13.5.5.2 Heat Wave

In 2020, California experienced a record high temperature from August to September, due to the heat wave effect. For example, on 16 August 2020, Death Valley National Park recorded 54.4°C and on 6 September 2020, 49.4°C in Los Angeles (Concepción de and Schwartz, 2020).

SGLI onboard GCOM-C can obtain various information by observing 19 wavelengths from near-ultraviolet region to thermal infrared region. Figure 13.6 is an image of LST estimated from an infrared band (wavelength 10.8, 12.0 μm) observed during daylight from 8 to 14 August 2020. LSTs of bare land, fields around forests, and shrubby areas exceeded 60°C. Based on the findings in the previous section that the plants were under extremely dry condition at that moment, it could be assumed that the fire could very easily occur.

13.5.5.3 Visualisation of Wildfires

JAXA's EO satellites can visualise phenomena on the ground, and in this case the background and effect of wildfires were visualised using both optical sensors and synthetic aperture radar, making the best use of their respective strengths. Optical

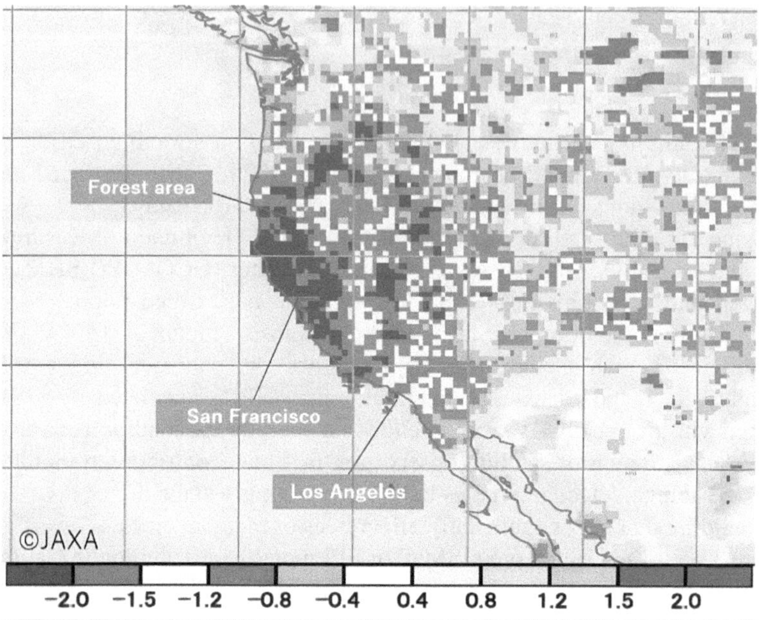

Figure 13.5 Example of Standardised Precipitation Index (SPI) (GSMaP) in a month. SPI of California calculated by GSMaP precipitation amount for February 2020. SPI in most part of California indicated less than −1.5, showing that California faced significant drought condition compared with normal year.

Source: JAXA (2020).

sensors onboard GCOM-C and GOSAT captured the condition and change of visible fires, as shown in Figures 13.7 and 13.8. Phased Array Type L-band Synthetic Aperture Radar-2 (PALSAR-2) onboard Advanced Land Observing Satellite-2 DAICHI-2 (ALOS-2) can observe ground surface day and night. By utilising electric waves penetrating clouds and smokes, the observation is not affected by the weather, which is not possible by optical sensors.

Figure 13.7 visualised California wildfires on 9 September 2020, by combining a red–green–blue (RGB) image of the GCOM-C visible band. The overlaid plotted red points show the "hotspots", the point which seems to be burning at the moment. The pixel exceeding a certain threshold value of brightness temperature was extracted as hotspots during the night without solar insolation using the thermal infrared band of GCOM-C from August 1 to 9 September 2020. The smoke by forest fires originating from the hot spots is observed, with the emission reaching far to the Pacific side.

Figure 13.8 shows the burning areas as of 3 September 2020, indicated in red points, extracting the difference before and after the fires from the images observed

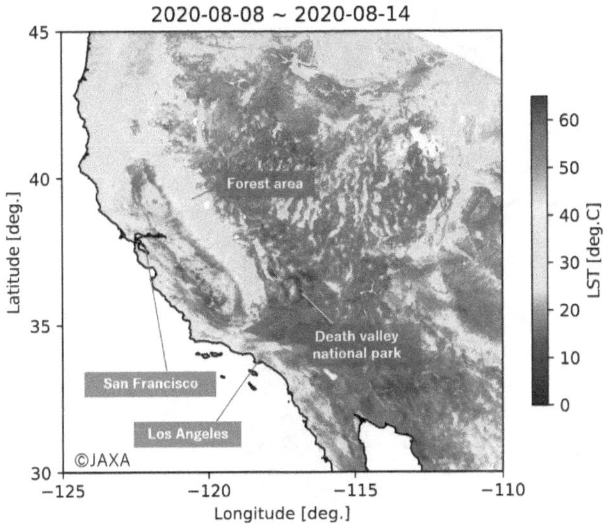

Figure 13.6 Average land surface temperature (LST) during 8–14 August 2020, observed by GCOM-C.

Source: JAXA (2020).

by Cloud and Aerosol Imager (CAI) onboard GOSAT. Figure 13.9 shows the comparison between visible image and ALOS-2/PALSAR-2 image on 1 September 2020. GCOM-C could not catch the change of grand surface because of the smoke covering the area. In such a case, PALSAR-2 can be used to clearly extract the forest burning area by merging the before and after images by HV polarised wave, detecting forest distribution indicated in red area in Figure 13.9 (right).

Thus, as seen in this example, drought and heatwave, which are the two main factors that could cause wildfires that may affect cities such as San Francisco and Los Angeles, could be observed by EO satellites. The visualisation of wildfires could assist stakeholders to take necessary actions to prevent negative effects on people living in the area, including habitants in populated cities.

13.6 Earth Digital Twin

While several case studies were demonstrated in the previous sections, technological innovation is needed to understand the whole Earth. It is in this context that the Earth digital twin is expected to be particularly effective in monitoring and predicting the Earth environment. There are movements towards this development in the public and private sectors of several countries (Bauer, Stevens and Hazeleger, 2021; Hoffman *et al.*, 2023).

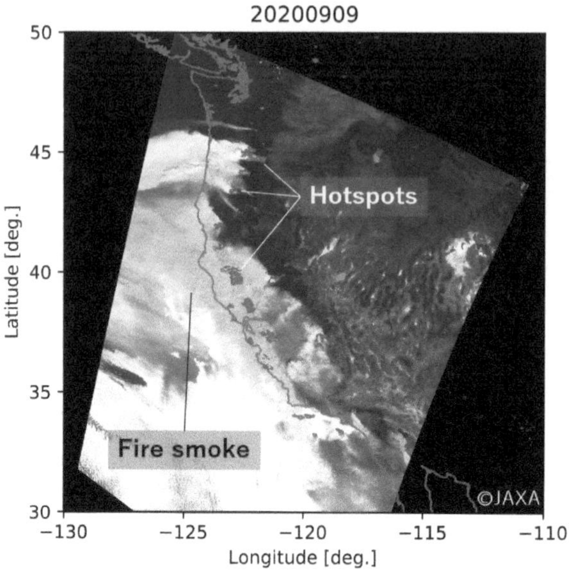

Figure 13.7 Visible images and hotspots in the west coast of USA on 9 September 2020, observed by GCOM-C. Originating from the hot spots, fire smoke by forest fires (indicated in the figure) is observed, reaching to the Pacific side.

Source: JAXA (2020).

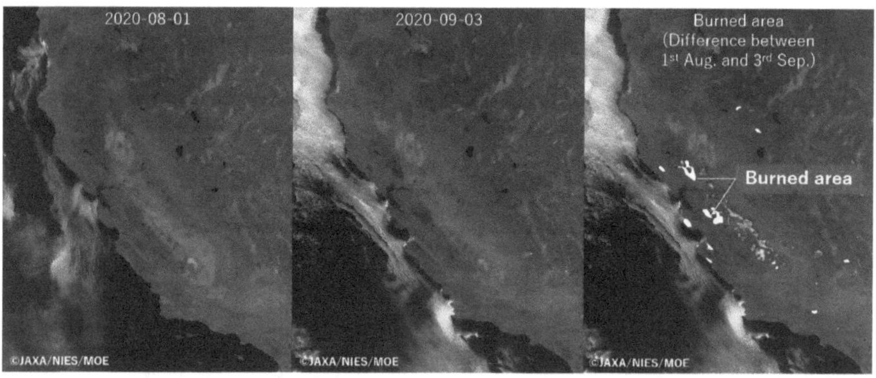

Figure 13.8 RGB composite images of the west coast observed by GOSAT. *From the left*: Before huge fires captured on 1 August 2020; after fire expansion captured on 3 September 2020; burning area as of 3 September 2020 indicated in red assumed by difference of two preceding images. For composite images, each data of 674, 870, 380 nm was allocated to RGB.

Source: JAXA (2020).

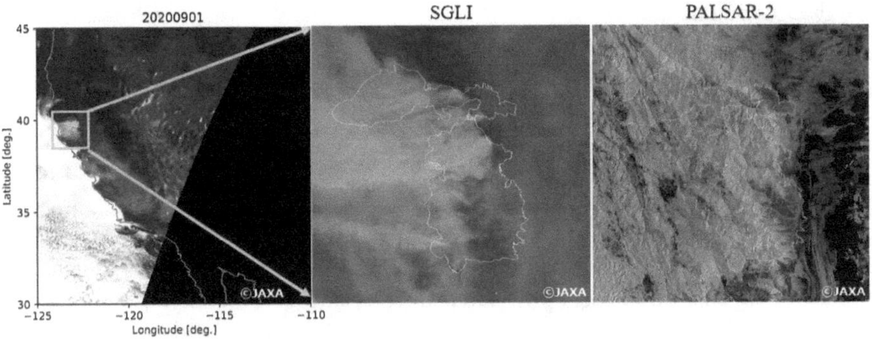

Figure 13.9 Comparison between visible image and PALSAR-2 image of the fire area on 1 September 2020. *Left*: Visible image of the west coast of North America captured by GCOM-C. *Middle*: Extended figure of the fire area (visible image) captured by GCOM-C. The white line shows the fire spread area released by the National Interagency Fire Centre up to September 10. *Right*: Extended figure of the fire area (PALSAR-2 image) captured by ALOS-2. RGB composite image of two terms, 7 July and 1 September 2020. (Red: HV image, before fire Green and Blue: HV image, after fire). The colour looks red at the fired area (see original figure) because backscattering coefficient decreased. The white line is the same as described in the *middle* image. PALSAR-2 image can show the damage condition of ground surface clearly, regardless of the smoke covering the area.

Source: JAXA (2020).

13.6.1 Earth Digital Twin: DestinE

A digital twin is a system that provides a twin in a digital space that is paired with the real world, making monitoring and simulation possible. In recent years, four-dimensional information (three-dimensional plus time) provided by the Earth digital twin has been used to provide information from the past to the future on water, food, energy, public health, etc., which is essential for socio-economic policies and industrial activities, and to create a new society. Efforts are underway to realise this.

There are descriptions of the development of digital twins of Earth funded by the European Union for its green transition, the Destination Earth (Bauer, Stevens and Hazeleger, 2021; Hoffman *et al.*, 2023). Hoffman describes that the European Destination Earth (DestinE) initiative aims at implementing a digital twin of the Earth system (Hoffman *et al.*, 2023). DestinE combines high-end physical with impact science enabled by novel digital technologies to create a new, interactive digital information system in support of decision-making.

Bauer pointed out that the digital twin of Earth creates an entirely new way for interacting with data and extracting the information that supports decision-making (Bauer, Stevens and Hazeleger, 2021), enabled by a digital platform that allows users to configure workflows for simulations, simulation–observation fusion, and add their own data and tools.

13.6.2 United States and Earth Digital Twin

As for the United States' perspective, NASA's Advanced Information Systems Technology Program defines an Earth System Digital Twin as composed of three components (Le Moigne and Smith, 2022):

1. A digital replica, that is an integrated picture of the past and current states of Earth systems, based on up-to-date, diverse, and continuous observations as well as on historical records;
2. Forecasting capabilities, providing an integrated picture of how Earth systems will evolve in the future from the current state,
3. Impact assessment capabilities, providing an integrated picture of how Earth systems could evolve under different hypothetical what–if scenarios, and how they could be impacted by other Earth systems as well as by human systems and activity.

13.6.3 Japanese Earth Digital Twin

Earth observation satellite data is expected to be used in society in areas such as national land planning and urban planning through digital twins that are fused with high-resolution models. Various industries, including agriculture, have been affected by climate change. Improvement and sophistication of prediction models are essential to countermeasures (mitigation and adaptation) to climate change and to improve the accuracy of predicting weather disasters that are becoming more frequent and severe.

To this end, it is important to strengthen four-dimensional information (three dimensions plus time changes). There is a need to acquire new technologies related to the collection of vertical information, whether on land or in the atmosphere, and there is a need to further utilise the research and development of 3D observation held by satellite organisations and the satellite capabilities held by domestic, international, and private businesses. It is expected that the temporal resolution will be enhanced. The Earth digital twin, which is made up of these satellite data and models, is expected to be used to improve operational efficiency in national and urban planning, as well as to analyse and simulate climate change and global environmental issues.

In Japan, data is generated by an active weather forecasting system that assimilates Earth satellite observation data into high-resolution numerical models, and by JAXA's system named as Realtime Weather Watch (NEXRA). There is a possibility that this data will contribute to Digital Earth and be used socially as a digital twin of the Earth. As the use of EO satellite data can be expected to expand through such initiatives, it is necessary for satellite data to adapt to trends in the use of data through new digital technologies. Japan should also consciously promote the construction of a global digital twin system that combines EO satellite data and high-resolution models and promote the development of a data and information base for solving global issues such as climate change.

13.7 Fusion of Earth Digital Twins and City Digital Twins for Energy Transition

It is assumed that energy transition will be encouraged and promoted by understanding and predicting human activities in the city and its effects accurately, by building smart cities underpinned by both city and Earth digital twins. This section studies such possibilities by examples of fusion of city and Earth digital twins.

13.7.1 Understanding Water Cycle to Contribute to Water Resource Management and Reducing Water-related Hazards

Monitoring and possibility prediction of global water resource distribution would assist decision-making in socio-economic activities. As already raised in the case of visualising wildfires, JAXA has been developing GSMaP for monitoring global water cycles.

GSMaP is the Japanese precipitation product, and graphical user interface of the JAXA Global Rainfall Watch website (JAXA, 2024) is available based upon the GSMaP product (Figure 13.10).

Impacts of climate change are obvious in local floods and drought. Against such a background, risk identification utilising long-term prediction before disasters is becoming increasingly important.

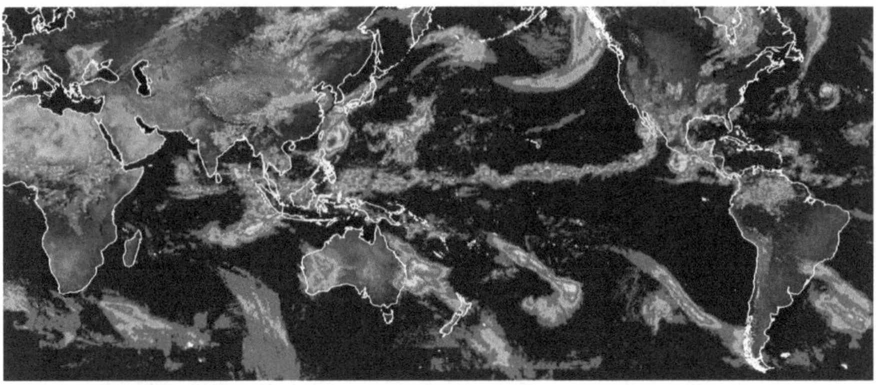

Figure 13.10 Example image of hourly global rainfall map by GSMaP.

Source: Extracted by JAXA, EORC.

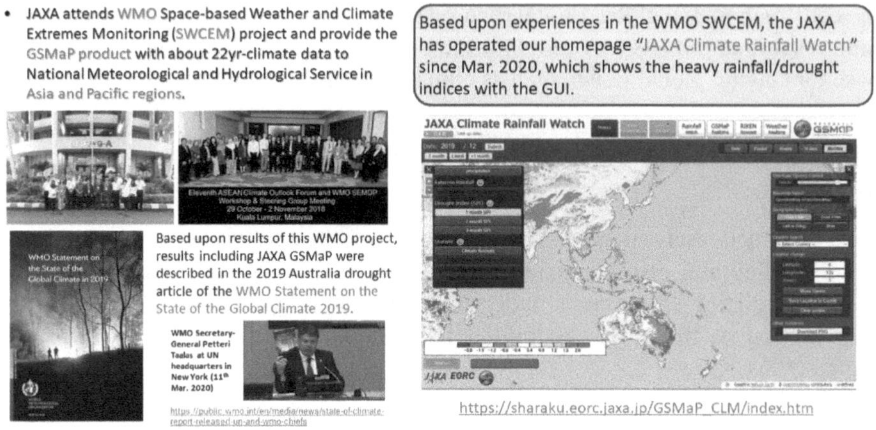

Figure 13.11 JAXA's monitoring of weather and climate as of 2023.

Source: Oki (2023).

In this regard, JAXA has been a member of the World Meteorological Organization's Space-based Weather and Climate Extremes Monitoring (SWCEM) project and provides the GSMaP product with about 22-year climate data to the National Meteorological and Hydrological Service in Asia and Pacific regions (Figure 13.11).

Because these observations are all at global level, it is necessary to downscale for local- and city level aiming to contribute to the assessment for impacts on human activities. In previous research, a downscaling method for climate model simulations using machine learning has been proposed (e.g. Yoshikane and Yoshimura ., 2023). JAXA is collaborating with universities and other organisations that are promoting such efforts and is also exploring the application of such methods to global satellite data products.

13.7.2 Monitoring Ecosystem and Carbon Cycle for Future Land Use Scenario

It is important to understand land cover to monitor carbon cycle by data–model fusion, and to reduce climate change uncertainties associated with land cover. Using the satellite data, JAXA has created the high-resolution land cover classification maps. The condition of land and the status of utilisation are classified into 12 categories: water, urban, rice paddy, crops, grassland, deciduous broadleaf forest (DBF), deciduous needle leaf forest (DNF), evergreen broadleaf forest (EBF), evergreen needle leaf forest (ENF), bare land, bamboo, and solar panel (JAXA, n.d.; Figure 13.12).

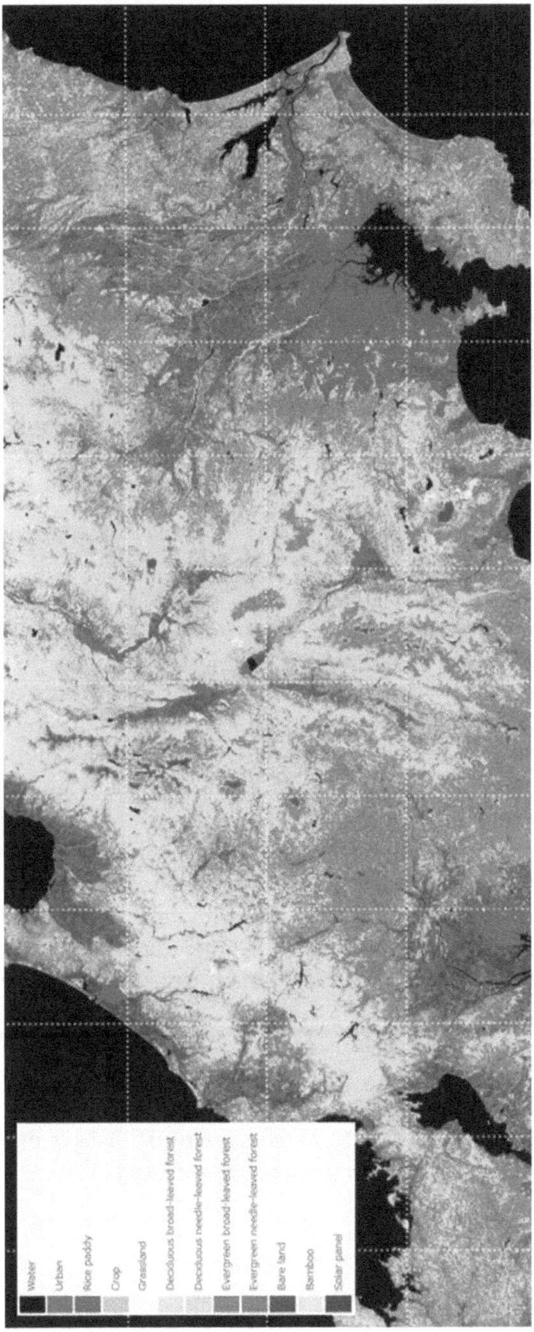

Figure 13.12 Example of land cover classification.

Source: JAXA (n.d.).

The maps will serve as basic information for various applications for the conservation of regions and lands such as ecosystem evaluation (such as distribution and habitat of animals and plants, various ecosystem services), resource management (such as agriculture, forestry and fisheries, landscape), and disaster countermeasures (such as flood and sediment disaster).

In addition, it would be essential for protecting lives and economic activities to understand intensifying geographical risks of natural disasters and prevent economic loss, thus sustainably coexisting with the environment. There is a shared awareness in the international community to disclose information to evade business risks, such as initiatives by the Task Force on Climate-related Financial Disclosures (TCFD) and the Task force on Nature-related Financial Disclosures (TNFD). Creating more high-resolution land use maps as part of the city digital twin would assist such efforts.

13.7.3 Marine Environment Monitoring for Sustainable Marine Resource Management and Environmental Conservation

Smart agriculture, forestry, and fishery would also have large impacts on urban life. As an example of areas that are not connected with a city digital twin but may be effective through analysis using Earth digital twin because of its large impact in urban life is marine-related area. Recently, climate change has caused many problems such as ocean warming, expansion of marine pollution, depletion of fishery resources, and decreased biodiversity.

SGLI onboard GCOM-C is conducting global observation for sea surface temperature (SST) and phytoplankton levels. With the observation of wide range at high frequency by SGLI will contribute to assessing the status of these problems, determining their causes, mitigating the damages, and resolving the issues. With its wide range of wavelengths from near ultraviolet and visible to thermal infrared, SGLI can capture reflected sunlight, ocean colour, underwater using a sensor. This coverage of the vast sea surface, which constitutes more than 70% of the Earth, is a possible contribution by satellite observation. With the 250 m resolution of SGLI, contribution is expected for conservation by monitoring, especially in coastal areas with large populations where land, ocean, and atmosphere are complexly interconnected. (JAXA, 2022)

GCOM-C high spatial resolution is expected to be able to observe ocean phenomena on smaller scales that have been difficult to do with previous sensors.

Climate change also affects island countries, maritime nations, and port cities. Island countries are increasingly becoming aware of the danger of sea level rise, and satellite observations are helpful in monitoring coastal erosion.

13.8 Conclusion

This chapter compiled and highlighted recent achievements of satellite EO data and its possibility to be utilised for the energy transition in smart cities. Satellite EO data alone may not necessarily be useful, however its characteristics and strengths combined with other data sources and integrated in data platform, will likely generate useful information, by providing data of the environment affecting lives in cities, from GHG to temperature, rain, wildfire, marine environment. Accurate information shall be the basis of decision-making.

In a future society with real-time simulation of wind by observation and model in the 3D city terrain model, safety operation of flying vehicles, and rulemaking of operation should be possible. Furthermore, with the digitalisation of the environment information should increase the safety of self-driving cars and ships, realising complete automation of aquaculture and agriculture. With the integration of various information by digital twin and AI on TOP, it will significantly lower the entry barrier to primary industry by providing necessary information and advice. Thus, optimisation in various issues related to resources, food supply, and transportation that are indispensable for city life would become a reality.

Regarding energy for the future smart city, the difference between "smart" and conventional city is that in the latter stable provision of energy has been the largest issue, whereas in the future smart city, management of cost, optimal and efficient allocation of energy would be a prerequisite. This implies that the energy sector should be a service sector with data innovation (Kashiwagi, 2019, p. 27).

While actual application should be worked out with relevant stakeholders, by realising the Earth Digital Twin and fusing with the city digital twin, a more accurate future projection would be possible. Use of emerging technologies such as generative AI and twin engines is also a trend that should encourage this transition. It would take both human activities and natural phenomena into account, thus contributing to transition and efficient use of energy.

Acknowledgements

The authors would like to thank Dr Riko Oki, Director of Earth Observation Research Centre, for her comments.

References

Anderson, K. *et al.* (2017) "Earth Observation in Service of the 2030 Agenda for Sustainable Development", *Geo-spatial Information Science*, 20(2), pp. 77–96.

Bauer, P., Stevens, B. and Hazeleger, W. (2021) "A Digital Twin of Earth for the Green Transition", *Nature Climate Change*, 11, pp. 80–83.

Cabinet Office, Government of Japan (2023) "スマートシティガイドブック第2版" [Smart City Guidebook 2nd edition], p.10. Available at: www8.cao.go.jp/cstp/society5_0/smartcity/guidebook.html. (Accessed: 23 March 2024).

Cabinet Office, Government of Japan (2024) "スマートシティとは" [What Is a Smart City?]. Available at: www8.cao.go.jp/cstp/society5_0/smartcity/index.html (Accessed: 23 March 2024).

Centre for Climate and Energy Solutions (2024) "Wildfires and Climate Change". Available at: www.c2es.org/content/wildfires-and-climate-change/ (Accessed: 23 March 2024).

Cheng, Y. *et al.* (2021) "Enhanced Simulation of an Asian Dust Storm by Assimilating GCOM-C Observations", *Remote Sensing*, 13, p. 3020.

Concepción de, L. and Schwartz, J. (2020) "Death Valley Just Recorded the Hottest Place on Earth". *The New York Times*, 17 August 2020. Available at: www.nytimes.com/2020/08/17/climate/death-valley-hottest-temperature-on-earth.html (Accessed: 23 March 2024).

Consortium for Satellite Earth Observation (CONSEO) (2024) "衛星地球観測コンソーシアム(CONSEO)による宇宙技術戦略へのインプットについて" [Inputs to the Space Technology Strategy by the Consortium for Satellite Earth Observation (CONSEO)]. Available at: www8.cao.go.jp/space/comittee/02-jissyou/jissyou-dai27/siryou2.pdf (Accessed: 23 March 2024).

Earth Observation Research Centre and Japan Aerospace Exploration Agency (EORC and JAXA) (2024a) "JAXA/GOSAT ANA Space Project Greenhouse Gas Observations of Biospheric and Local Emissions from the Upper Sky". Available at: www.eorc.jaxa.jp/GOSAT/ANAexp/index_e.html (Accessed: 23 March 2024).

Earth Observation Research Centre and Japan Aerospace Exploration Agency (EORC and JAXA) (2024b), "Flight Map". Available at: www.eorc.jaxa.jp/GOSAT/ANAexp/flightmap.html. (Accessed: 23 March 2024).

Earth Observation Research Centre and Japan Aerospace Exploration Agency (EORC and JAXA) (2024c) "What Is Today's Earth?". Available at: www.eorc.jaxa.jp/water/about.html (Accessed: 23 March 2024).

Hoffmann, J. *et al.* (2023) "Destination Earth – A Digital Twin in Support of Climate Services", *Climate Services*, 30, p. 100394.

Intergovernmental Panel on Climate Change (IPCC) (n.d.) "Weather and Climate Extreme Events in a Changing Climate", in *Sixth Assessment Report, Working Group 1: The Physical Science Basis*. Available at: www.ipcc.ch/report/ar6/wg1/chapter/chapter-11/ (Accessed: 23 March 2024).

Japan Aerospace Exploration Agency (JAXA) (2018) "「しきさい」が捉えた日本の猛暑" [Japan's heatwave captured by 'SHIKISAI']. Available at: https://earth.jaxa.jp/ja/earthview/2018/08/01/1381/index.html. (Accessed: 23 March 2024).

Japan Aerospace Exploration Agency (JAXA) (2020) "Visualization of California Wildfires by Earth Observatory Satellites". Available at: https://earth.jaxa.jp/en/earthview/2020/09/28/1631/index.html (Accessed: 23 March 2024).

Japan Aerospace Exploration Agency (JAXA) (2022) " 'SHIKISAI' (GCOM-C) Marks the Fifth Year Since Its Successful Launch: Achievements in Ocean Environment". Available at: https://earth.jaxa.jp/en/earthview/2022/07/11/7123/index.html (Accessed: 23 March 2024).

Japan Aerospace Exploration Agency (JAXA) (2024) "JAXA Global Rainfall Watch". Available at: https://sharaku.eorc.jaxa.jp/GSMaP/index.htm (Accessed: 23 March 2024).

Japan Aerospace Exploration Agency (JAXA) (n.d.), "Land Cover Data". Available at: https://earth.jaxa.jp/ja/data/products/land-cover/index.html (Accessed: 23 March 2024).

Japan Meteorological Agency (JMA) (2024) "Aeolian Dust Information". Available at: www.data.jma.go.jp/gmd/env/kosa/fcst/en/ (Accessed: 23 March 2024).

Kashiwagi, T. (2019) "エネルギー変革を見据えてスマートシティを推進" [Promoting Smart Cities in Anticipation of Energy Transformation], in H. Ishida and T. Kashiwagi (eds), *スマートシティ Society5.0の社会実装* [Smart City: Social Implementation of Society 5.0]. Tokyo: Jihyosha, pp. 18–27.

Le Moigne, J. and Smith, B. (2022) "NASA Earth Science Technology Office, Advanced Information Systems Technology (AIST) Earth Systems Digital Twin (ESDT)". Available at: https://esto.nasa.gov/files/ESDT_Workshop_Report.pdf (Accessed: 23 March 2024).

Ma, W. *et al.* (2021) "Applicability of a Nationwide Flood Forecasting System for Typhoon Hagibis 2019", *Scientific Reports*, 11, p. 10213.

Ministry of Land, Infrastructure, Transport and Tourism (MLIT) (2024) "PLATEAU". Available at: www.mlit.go.jp/plateau/ (Accessed: 23 March 2024).

Nasahara, K. (2022) "3. Monitoring Land Surface Temperature at 250 m Resolution". Available at: https://earth.jaxa.jp/en/earthview/2022/06/14/7032/index.html (Accessed: 23 March 2024).

National Integrated Drought Information System (NIDIS) (2021) "Drought Status Update and 2020 Recap for California–Nevada". Available at: www.drought.gov/drought-status-updates/drought-status-update-and-2020-recap-california-nevada (Accessed: 23 March 2024).

National Research Institute for Earth Science and Disaster Resilience (2024) "Shared Information Platform for Disaster Management". Available at: www.sip4d.jp/ (Accessed: 23 March 2024).

Oki, R. (2023) "JAXA's Activities on Water Cycle Variation and Climate Change Studies", CEOS/CGMS WG Climate 18, 28 February 2023, Tokyo, Japan. Available at: https://ceos.org/document_management/Working_Groups/WGClimate/Meetings/WGClimate-18/Presentations/20230228_WGClimate_Oki_fin.pdf (Accessed: 23 March 2024).

Sugita, N. *et al.* (2023) "Earth Observing Dashboard for Societal Benefits: The Development of Trilateral Collaboration and Beyond. Global Space Conference on Climate Change (GLOC)", 23–25 May 2023, Oslo.

World Economic Forum (2017) "Every Person in London Now Breathes Dangerous Levels of Toxic Air". Available at: www.weforum.org/agenda/2017/10/london-dangerous-levels-of-toxic-air-pollution/ (Accessed: 23 March 2024).

World Health Organization (WHO) (2023) "Monitoring Air Pollution Levels Is Key to Adopting and Implementing WHO's Global Air Quality Guidelines". Available at: www.who.int/news/item/10-10-2023-monitoring-air-pollution-levels-is-key-to-adopting-and-implementing-who-s-global-air-quality-guidelines#:~:text=Environmental%20risks%20cause%2012%25%20of,lower%20concentrations%20than%20previously%20recognized (Accessed: 23 March 2024).

World Health Organization (WHO) (2024) "Air Pollution: Overview". Available at: www.who.int/health-topics/air-pollution#tab=tab_1 (Accessed: 23 March 2024).

World Health Organization (WHO) (n.d.) "Wildfires". Available at: www.who.int/health-topics/wildfires#tab=tab_1 (Accessed: 23 March 2024).

World Health Organization Africa (WHO Africa) (n.d.). "Overview (Air Pollution)". Available at: www.afro.who.int/node/5526 (Accessed: 23 March 2024).

Yoshida, M. *et al.* (2018) "Common Retrieval of Aerosol Properties for Imaging Satellite Sensors", *Journal of the Meteorological Society of Japan*, 96B, pp. 193–209.

Yoshida, M. *et al.* (2021) "Retrieval of Aerosol Combined with Assimilated Forecast", *Atmospheric Chemistry and Physics*, 21, p. 1797.

Yoshikane, T. and Yoshimura, K. (2023) "A Downscaling and Bias Correction Method for Climate Model Ensemble Simulations of Local-Scale Hourly Precipitation", *Scientific Reports* 13, p. 9412.

Yoshimura, K. *et al.* (2008) "Toward Flood Risk Prediction: A Statistical Approach Using a 29-Year River Discharge Simulation over Japan", *Hydrological Research Letters*, 2, pp. 22–26.

Yumimoto, K. *et al.* (2017) "JRAero: The Japanese Reanalysis for Aerosol v1.0", *Geoscientific Model Development*, 10, pp. 3225–3253.

Yumimoto, K. *et al.* (2018) "Assimilation and Forecasting Experiment for Heavy Siberian Wildfire Smoke in May 2016 with Himawari–8 Aerosol Optical Thickness", *Journal of the Meteorological Society of Japan*, 96B, pp. 133–149.

Index

Note: Endnotes are indicated by the page number followed by 'n' and the endnote number e.g., 20n1 refers to endnote 1 on page 20.

Kolibri 360 GmbH
Ferdinand-Jühlke-Straße 7
99095 Erfurt, Deutschland
produktsicherheit@kolibri360.de